# MARS
## UNMASKED
### The Changing Face of Urban Operations

Sean J. A. Edwards

Prepared for the
**United States Army**

## RAND
**ARROYO CENTER**

For more information on the RAND Arroyo Center, contact the Director of Operations, (310) 393-0411, extension 6500, or visit the Arroyo Center's Web site at http://www.rand.org/organization/ard/

This monograph is a case study analysis of three recent urban operations. The objective of this research was to update lessons learned about military urban operations and determine the significance of recent changes. The three cases examined—Panama in 1989, Somalia in 1992–1993, and Chechnya in 1994–1996—captured the range of political constraints that military forces must operate under in urban environments.

This research was conducted for the project "Military Operations on Urbanized Terrain," directed by Russell Glenn and Randy Steeb. The project was sponsored by the Office of the Deputy Assistant Secretary of the Army for Research and Technology and conducted in the Force Development and Technology Program of RAND's Arroyo Center, a federally funded research and development center sponsored by the United States Army. This research should be of interest to anyone concerned about recent trends in the conduct of urban operations.

# CONTENTS

# FIGURES

# TABLES

The likelihood that U.S. military forces will be called on to fight in cities is increasing. There are many reasons for this trend: continued urbanization and population growth; a new, post–Cold War U.S. focus on support and stability operations; and a number of new political and technological incentives for U.S. adversaries to resort to urban warfare.

For instance, urban warfare is thought by some adversaries to be a useful asymmetric approach to fighting the U.S. military. They believe the American public has an antiseptic idea of war, an unrealistic expectation that it can be waged with minimal casualties. In this view, such sensitivity becomes an Achilles heel because inflicting a sufficient number of American casualties has the potential to undermine U.S. domestic political support. Cities offer physical cover—three-dimensional urban terrain—and political cover—the more stringent rules of engagement (ROE) associated with the presence of noncombatants. Both types of cover limit the effectiveness of U.S. heavy weapons such as tanks, artillery, and airpower. Weaker opponents can use cities to avoid heavy weapons, leverage the noncombatant population, and "even" the odds by fighting infantry-versus-infantry battles only.

If urban warfare is more likely to occur, it is imperative that U.S. military forces be ready to fight in cities. At the present time, U.S. doctrine on urban operations—also called Military Operations on

Urbanized Terrain (MOUT)—is based in part on historical case studies that occurred seventeen years or more in the past.[1]

Lessons that predate the early 1980s may be irrelevant or less important today, especially because of the larger number of political considerations that have restricted the use of force in more recent urban operations. U.S. MOUT doctrine requires an update that accounts for lessons from the last ten years. This monograph is an exploratory case study analysis of three recent urban operations: U.S. peace operations in Somalia (specifically, the Mogadishu firefight of October 3–4, 1993); Operation Just Cause (specifically, the urban battles that occurred in cities, towns, airports, and bases throughout Panama in December 1989); and the Chechen War (specifically, the battles for Grozny in December 1994–February 1995 and August 1996).

These three cases are similar to MOUT studied in the past in that combat occurred and in that the essential unit of analysis—the urban battle—remains the same.

The Mogadishu firefight started with a special-purpose raid by a company-sized element of U.S. commandos to abduct hostages from Mohammed Aideed's Somali clan. The mission went awry after two Blackhawk helicopters were shot down. Thousands of Somali guerrillas and civilians swarmed around the embattled U.S. commandos and convoys sent in to rescue them. Eighteen Americans were killed.

Operation Just Cause was a joint operation by over 26,000 U.S. troops to attack 27 objectives throughout Panama, including enemy troop concentrations and airports, and media, transportation, and command nodes. U.S. forces were ordered to overthrow the dictatorship of Manuel Antonio Noriega. Urban combat occurred during airfield seizures and deliberate attacks on Panamanian Defense Force (PDF) positions in military bases and cities. Most fighting was over within a couple of days. Twenty-three American soldiers were killed.

---

[1]In *Modern Experience in City Combat* (1987), R. D. McLaurin et al. distilled lessons from 22 urban conflicts ranging from Stalingrad (1942) to Beirut (1982). This particular report served as the basis for historical lessons that appeared in the official manual, Department of the Navy, *Military Operations on Urbanized Terrain (MOUT)*, Marine Corps Warfighting Publication 3-35.3, Washington, D.C.: U.S. Government Printing Office, 1998.

The Chechen War was fought between Chechen insurgents seeking independence and the Russian army, air force, and internal security forces. This two-year guerrilla war ran the gamut of urban operations, from small-scale Chechen raids into the Russian cities of Budyonnovsk and Kizlyar-Pervomaiskoye to high-intensity MOUT within the city of Grozny. The two major battles for Grozny involved tens of thousands of Russian soldiers and hundreds of tanks. Over 6,000 Russian soldiers were killed overall.

The primary objective of this monograph is to determine whether recent changes in the nature of urban operations are significant and to identify any policy implications for U.S. Army doctrine.

The main findings of this research are as follows:

- Several important elements of urban operations that previous studies have identified—such as situational awareness, intelligence, airpower, surprise, technology, combined arms, and joint operations—are no more decisive today than they were in the past.

- In the last decade, technological, social, and political changes have caused the following MOUT elements to become relatively more significant: the presence of the media, the presence of noncombatants, ROE, and information operation tools such as psychological operations (PSYOP), public affairs (PA), civil affairs (CA), and political-military strategy.

- Information technology, recent historical precedents, asymmetric responses, and shifting political justifications for the use of force have combined to exacerbate a longstanding geostrategic problem for conventional powers: how to wage restricted urban warfare while keeping casualties below some threshold of public tolerance.

- Recent trends indicate that urban operations should focus more on information-related factors that manipulate the will of the opposing population. This is not to say that information-related factors such as PSYOP or public affairs are more decisive than a "traditional" MOUT factor like airpower or combined arms teams. Killing the enemy's troops will probably remain the most efficacious way to defeat his will to fight. However, the marginal return from leveraging an information factor—such as the media—may be greater than the marginal return of applying more firepower.

## THE CHANGING FACE OF URBAN OPERATIONS

Political, technological, and social developments appear to be changing the way democratic nations justify and conduct urban operations. These developments—which include the spread of information technology, the growing presence of the media and noncombatants, changing standards of morality, and the increasing number of humanitarian operations, insurgencies, and asymmetric responses by weaker opponents—have increased the importance of information operations (and related activities). Information operations focus on the perception and will of the people fighting the war: the support of both the domestic population at home and the support of the indigenous population in the urban operations theater. Recent urban operations reinforce the notion that winning a conflict is about subduing the will of the enemy through information operations as well as destroying his military forces. To use a mythological metaphor, Mars, the Roman god of war, needs to be unmasked to reveal how warfare is increasingly waged in both the physical and informational realms.[2]

In the last decade, the political environment behind urban operations changed in several ways. For the United States, military operations were characterized by greater concern over public opinion, casualties of all sorts (including friendly, noncombatant, and even enemy casualties), and humanitarian issues. Because military action was justified for moral or humanitarian reasons, it was important for U.S. forces to gain the moral high ground. This more altruistic concept of national interest has been called "the Clinton Doctrine." When military action was conducted for less-than-vital national security threats, political support at home was more fragile and sus-

---

[2]Some authors believe that the Greek god Athena is the metaphor most appropriate for the information age. See John Arquilla and David Ronfeldt (eds.), *In Athena's Camp: Preparing for Conflict in the Information Age,* Santa Monica, CA: RAND, MR-880-OSD/RC, 1997.

ceptible not only to casualties, but also to enemy information operations.

Information technology increased the reach and responsiveness of the media. News reporters were present on the battlefield in greater numbers. The growth of media technology capable of recording battlefield drama introduced new political constraints on the use of military force. Democratically elected leaders were loath to expose voters to the brutal images of war.

These changes in the nature of urban operations have increased the significance of the media, noncombatants, ROE, information operation tools such as psychological operations (PSYOP), public affairs (PA), civil affairs (CA), and political-military strategy. Many of these elements are synergistic, so a successful political-military strategy must integrate information operation tools (PSYOP, PA, and CA) with the media. For example, public affairs, civil affairs, and psychological operations can help manage the perception of people within the area of operations. PSYOP and civil affairs units help remove noncombatants before a battle commences (thereby lowering possible noncombatant casualties) and increase human intelligence (HUMINT). PA and CA units interact with the media. ROE affect PA, CA, and PSYOP. Permissive ROE can precipitate civilian casualties, which attract more media. Overly restrictive ROE can cause friendly casualties. Some ROE—like graduated response approaches that use loudspeakers, warning shots, and firepower demonstrations—have PSYOP implications.

Specific lessons from Panama, Somalia, and Chechnya help illustrate how these elements interact:

- The presence of noncombatants significantly affected tactics, planning, ROE, and political-military strategy. Noncombatants were present in greater numbers, they played an active role in the fighting, they made ROE more restrictive, and they attracted the media.

- Balancing ROE proved to be difficult, especially in the high-intensity case. Constructing and managing flexible ROE that were neither restrictive nor permissive was critical. When improper ROE resulted in excessive civilian deaths and collateral damage, other MOUT elements such as the media and enemy IO could exploit the damage for their own interests. ROE also affected tactics and prevented the use of armor, artillery, and airpower on occasion. As a result, MOUT tactics, techniques, and procedures (TTPs) sometimes conformed more to a political logic than to a military logic (at least before excessive casualties begin to occur).

- The media was more significant because of the larger number of reporters and the portability of their information technology. It was easier for reporters to gain access to peace enforcement missions. All belligerents found the media a useful information tool for PSYOP, IO in general, civil affairs, and public affairs.

- PSYOP and civil affairs operations proved indispensable in influencing the will of the civilian populations involved. PSYOP were used to increase the number of noncombatants present. PSYOP were conducted by combining daring military raids with media exposure.

- The failure of political leadership to communicate the national interests at stake in Somalia and Chechnya lowered the public's threshold for casualties. It was important to have clear objectives before using military force, to avoid mission creep, and have a clear exit strategy. The lack of political leadership also had a corrosive effect on morale in one case.

Recent urban operations also showed that many elements of MOUT have not changed in any fundamental way. In particular:

- Complete situation awareness will remain an elusive goal for some time to come, just as it was in the past.[3] There were two reasons for this in our case studies—the unavailability of HUMINT and an inability to transmit sufficient information in the harsh electromagnetic conditions of the urban landscape.

- Airpower proved to be a mixed blessing in recent urban operations because of the presence of noncombatants, ROE, and capable air defense threats.[4] Urban terrain, poor weather, and an inability to precisely engage dispersed infantry with air-to-

---

[3]In fact, complete situational awareness may never be possible.

[4]For the purposes of this monograph, airpower includes rotary-wing aircraft.

ground munitions also contributed to the mixed performance of airpower. Airpower was effective in joint operations around the perimeter of small villages and towns that could be isolated, against specific strongpoints that could be pinpointed, and in open areas in clear weather. Helicopters were vulnerable in MOUT environments where dismounted infantry carrying man-portable surface-to-air (SAM) weapons could conceal themselves within crowds of noncombatants.

- Urban warfare technologies employed in the 1990s did not differ significantly from technologies available before 1982. Weapons remained essentially the same, especially when ROE prohibited the stronger side from fielding advanced tanks and artillery. Commercial-off-the-shelf (COTS) equipment, nonlethal weapons, and precision-guided munitions (PGMs) were either not used, not considered, or were not decisive.

- The advantage of surprise was critical to the outcome of all three case studies, but it was neither more nor less decisive than in the past.

- Combined arms teams were essential if friendly casualties needed to be minimized, but they also resulted in more collateral damage and noncombatant casualties. In the surgical and precision cases, combined arms teams were generally restricted by ROE.

- Command, control, and communication problems continued to plague joint operations. Communication between air and ground forces was a problem in all three case studies. Miscommunication between ground units and close air support (CAS) assets also caused some cases of fratricide.

When civilians are present in large numbers, their support may be the center of gravity, especially in insurgencies. Noncombatants can conceal the enemy, provide intelligence, and take an active role in the fighting. In this age of restricted warfare, the effort to subdue the will of the enemy requires a systems approach that combines information-related activities with the application of military force. Information-related activities such as civil affairs, public affairs, PSYOP, balanced ROE, and information operations in general can possibly offer higher marginal returns (for resources expended) when noncombatants are central to the urban operations.

In future conflicts, it should be anticipated that some U.S. adversaries will recognize the growing importance of these information elements and leverage them as part of an asymmetric response to American firepower. War has always been waged in both the physical and the informational realm, but the changes under way today make it imperative that we pay more attention to the latter.

# ACKNOWLEDGMENTS

I am grateful to many people for their advice and help on this monograph. Randy Steeb and John Matsumura extended their patience and support for this research effort. Michele Zanini made himself available to discuss several of my main themes. LTG Ron Christmas (USMC, ret.) and John Gordon contributed excellent technical reviews. Major Wayne Barefoot provided several pages of comments on how to improve my thesis. Sandy Berry offered some helpful suggestions regarding the methodology. Olya Oliker, Alan Vick, and Eric Larson were kind enough to provide me feedback. John MacDonald of Williamstown, Massachusetts created the beautiful cover for this monograph. Russell Glenn deserves special appreciation for serving as both a mentor and an editor throughout the writing of this work. Finally, Nikki Shacklett edited and improved the final product.

Needless to say, any error or omission is the author's responsibility alone.

# ABBREVIATIONS

| | |
|---|---|
| AAA | Anti-aircraft artillery |
| APC | Armored personnel carrier |
| C2 | Command and control |
| C2W | Command and control warfare |
| C3 | Command, control, and communications |
| C3I | C3 and intelligence |
| C4 | Command, control, communications, and computers |
| C4ISR | C4 and intelligence, surveillance and reconnaissance |
| CA | Civil affairs |
| CAS | Close air support |
| CINCSO | Commander in Chief, U.S. Southern Command |
| CJTF | Commander of the Joint Task Force |
| CNN | Cable News Network |
| CONUS | Continental United States |
| COTS | Commercial off-the-shelf |
| DoD | Department of Defense |
| EW | Electronic warfare |
| HMMWV | High Mobility Multi-Purpose Wheeled Vehicle |

| | |
|---|---|
| HRO | Humanitarian relief organization |
| HUMINT | Human intelligence |
| IFV | Infantry fighting vehicle |
| IO | Information operations |
| IW | Information warfare |
| JSOTF | Joint Special Operations Task Force |
| JSTARS | Joint Surveillance Target Attack Radar System |
| JTF | Joint Task Force |
| JTFSO | Joint Task Force South |
| LAW | Light anti-tank weapon |
| LIC | Low-intensity conflict |
| MANPADS | Man-Portable Air Defense System |
| MMW | Millimeter wave |
| MOUT | Military operations on urbanized terrain |
| MP | Military police |
| MTW | Major theater war |
| MVD | Ministry of Internal Affairs (Russia) |
| NATO | North Atlantic Treaty Organization |
| NCMD | North Caucasus Military District |
| NEO | Noncombatant evacuation operation |
| NGO | Nongovernmental organization |
| OAF | Operation Allied Force |
| OCJCS | Office of the Chairman of the Joint Chiefs of Staff |
| OJC | Operation Just Cause |
| OOTW | Operations other than war |
| OPSEC | Operational security |
| PA | Public affairs |
| PDF | Panamanian Defense Force |

| | |
|---|---|
| PGM | Precision-guided munition |
| PSYOP | Psychological operations |
| QRF | Quick Reaction Force |
| ROE | Rules of engagement |
| RPG | Rocket-propelled grenade |
| SAM | Surface-to-air missile |
| SEAL | Sea-Air-Land naval commandos |
| SIGINT | Signals intelligence |
| SIO | Special information operations |
| SITREP | Situation report |
| SNA | Somali National Alliance |
| SOUTHCOM | Southern Command |
| SSC | Small-scale contingency |
| SSO | Support and stability operations |
| SUV | Sport utility vehicle |
| TOC | Tactical operations center |
| TOW | Tube-launched, Optically-tracked, Wire-guided antitank missile |
| TTP | Tactics, techniques, and procedures |
| UAV | Unmanned aerial vehicle |
| UN | United Nations |
| UNITAF | Unified Task Force |
| UNOSOM | United Nations Operation in Somalia |
| USARSO | U.S. Army South |
| USCENTCOM | United States Central Command |
| USSR | Union of Soviet Socialist Republics |

# Chapter One

# INTRODUCTION

The likelihood that U.S. military forces will fight in cities is increasing. There are many reasons for this trend: continued urbanization and population growth; a new, post–Cold War U.S. focus on support and stability operations; and a number of new political and technological incentives for U.S. adversaries to resort to urban warfare.

As urbanization rapidly increases around the globe, urban conflict is also likely to rise. When rural populations migrate to cities, the guerrilla forces that depend on them for food, information, concealment, and general support must follow.[1] Press attention is easier to get in the city for terrorists seeking media exposure.

The realities of the post–Cold War security environment and the current goals of the U.S. National Security Strategy make it more likely that our military forces will operate in cities. The end of the Cold War has already produced a dramatic increase in U.S. deployments for peace, humanitarian assistance, and disaster relief operations. Between 1945 and 1989, the Army conducted two large peace operations—Dominican Republic and Egypt—and since then no fewer than six such operations (Iraq, Somalia, Haiti, Macedonia, Bosnia,

---

[1]Cities offer large pools of people for insurgency propaganda and recruitment. See Major Steven P. Goligowski, *Operational Art and Military Operations on Urbanized Terrain*, Fort Leavenworth, KS: School of Advanced Military Studies, U.S. Army Command and General Staff College, 1995, p. 5; and also Jennifer Taw and Bruce Hoffman, *The Urbanization of Insurgency: The Potential Challenge to U.S. Army Operations*, Santa Monica, CA: RAND, MR-398-A, September 1994, p. 7.

and the Sinai).[2]  At the same time, support and stability operations frequently take place in towns, villages, and cities because political, economic, and cultural nodes are located in urban areas.

There are several political incentives for U.S. adversaries to fight in cities. Many adversaries believe that the American public has an antiseptic view of war, an unrealistic expectation that it can be waged with minimal casualties. The memory of the clean "Nintendo" victory during the Persian Gulf War is still fresh. The recent victory of Operation Allied Force in Serbia was achieved without a single American combat casualty.[3]  Adversaries believe that the U.S. public's misplaced confidence in high-technology weapons increases our sensitivity to casualties. This sensitivity is viewed as an Achilles heel because the infliction of a sufficient number of American casualties has the potential to undermine domestic political support for military action. This is especially true for operations that the American public perceives as involving less-than-vital national interests.

Fighting in cities offers an adversary a way to inflict higher casualties.[4]  The presence of noncombatants in urban areas usually requires more stringent rules of engagement (ROE), which prohibit or limit the effectiveness of heavy weapons such as tanks, artillery, and airpower. Adversaries can use cities to avoid these heavy weapons and "even" the odds of facing U.S. military might by fighting infantry-versus-infantry battles.[5]

---

[2]See Jennifer Taw, David Persselin, and Maren Leed, *Meeting Peace Operations' Requirements While Maintaining MTW Readiness*, Santa Monica, CA: RAND, MR-921-A, 1998, p. 5.

[3]NATO's 33,000-sortie, 11-week air campaign scored a near-perfect safety record, with no pilot losses due to hostile fire. Columnist Bradley Graham wrote that the focus on minimizing casualties altered tactics and strategy. Minimizing casualties played a part in the decision to hold ground troops out of the fight, to rely on air strikes, and to keep attack planes at altitudes above 15,000 feet during the early weeks of the campaign, out of range of some Yugoslav air defenses. See Bradley Graham, "War Without 'Sacrifice' Worries Warriors," *The Washington Post*, June 29, 1999.

[4]Every major adversary of the United States since World War II has sought to maximize U.S. casualties and create a strategic psychological effect. See Steve Hosmer, "The Information Revolution and Psychological Effects," in Zalmay M. Khalilzad and John P. White (eds.), *Strategic Appraisal: The Changing Role of Information in Warfare*, Santa Monica, CA: RAND, MR-1016-AF, 1999, p. 224.

[5]Hypothetical adversaries in high-level wargames are already exploring the advantage of using an asymmetric strategy to fight future U.S. forces. In recent Army After Next

An enemy's desire to avoid U.S. heavy forces will only increase as the U.S. Army completes its Force XXI modernization to a more lethal force.[6] Because urban warfare is primarily an infantry fight, it is a form of warfare that lends itself least to the application of advanced technology.[7] Force XXI will not significantly increase the lethality of the average infantryman, and infantrymen remain vulnerable to enemy small arms fire. Force XXI modernization will increase the firepower of armor and artillery, but in many cases large-caliber cannons and rockets are not discriminating enough for the ROE typically exercised on the urban battlefield. Even when these weapons can be used, they require dismounted infantry protection.

On top of this, the main advantage that Force XXI modernization is seeking to provide—superior situational awareness—is seriously degraded by urban terrain. The radios, computers, sensors, and communications equipment that make up the Force XXI tactical internet use wireless transmissions to establish situational awareness among mobile users. Buildings, walls, and other obstructions in urban terrain reflect, absorb, and block communications signals.

Given the demographic trends, political incentives, and technological limitations described above, it is imperative that doctrine be sound. At the present time, U.S. doctrine on urban operations—also called military operations on urbanized terrain (MOUT)— is based in part on historical case studies that occurred 18 years or more in the

---

wargames, Red military forces sought to use urban terrain to fight blue forces. See Sean D. Naylor, "A Lack of City Smarts? War Game Shows Future Army Unprepared for Urban Fighting," *Army Times*, May 11, 1998.

[6]The U.S. Army is currently being modernized into Army XXI, a more lethal, digitized force. Heavy platforms benefit the most from this digitization, light forces less so. For example, airpower and heavy ground forces have added a plethora of precision-guided munitions (PGMs), enhanced sensors, and C4I systems, whereas dismounted light infantrymen have only so much muscle to carry their basic load. The benefits of standoff precision fire accrue mostly to those platforms with the communication and information capability necessary to call for real-time support.

[7]The most effective way to conduct MOUT is with combined arms teams of infantry, armor, and engineers, but in tactical situations tanks should primarily be used as a supporting arm for the infantry because of the threat of rocket-propelled grenades (RPGs). Tanks should be used to seal off city blocks, repel counterattacks, and provide covering fire along streets. During movement, tanks should move behind their own infantry at a distance beyond the range of enemy antitank weapons. See Timothy L. Thomas, "The Battle for Grozny: Deadly Classroom for Urban Combat," *Parameters*, Summer 1999.

past.[8] For example, the Marine manual *Military Operations on Urbanized Terrain (MOUT)* outlines lessons drawn from 22 battles in "modern" urban warfare history between 1943 and 1982. This is in part attributable to the MOUT manual's reliance on one particular report, *Modern Experience in City Combat,* by R. D. McLaurin et al. (1987).

Many authors are suggesting that it is time to update these lessons on MOUT by looking at more recent historical cases.[9] Lessons from urban operations that predate the early 1980s may be irrelevant or less important today.[10] Urban operations now seem more probable in missions other than conventional large-scale war—missions that

---

[8]Current MOUT doctrine can be found in three manuals: Department of the Army, *Military Operations on Urbanized Terrain,* Field Manual (FM) 90-10, Washington, D.C.: U.S. Government Printing Office, 1979; Department of the Army, *An Infantryman's Guide to Combat in Built-up Areas,* Field Manual (FM) 90-10-1, Washington, D.C.: U.S. Government Printing Office, 1993; Department of the Navy, *Military Operations on Urbanized Terrain (MOUT),* Marine Corps Warfighting Publication 3-35.3, Washington, D.C.: U.S. Government Printing Office, 1998.

[9]Several authors have identified a need to update various aspects of U.S. MOUT doctrine. Two reports by Russell Glenn, *Combat in Hell,* Santa Monica, CA: RAND, MR-780-A/DARPA, 1996, and *Marching Under Darkening Skies: The American Military and the Impending Urban Operations Threat,* Santa Monica, CA: RAND, MR-1007-A, 1998, both highlight many problems with the current state of U.S. MOUT doctrine. Other writers have also conducted historical analysis of MOUT cases after 1982 and called for appropriate doctrinal changes. They include Major Phillip T. Netherly, *Current MOUT Doctrine and Its Adequacy for Today's Army,* Fort Leavenworth, KS: School of Advanced Military Studies, U.S. Army Command and General Staff College, 1997; Major Robert E. Everson, *Standing at the Gates of the City: Operational Level Actions and Urban Warfare,* Fort Leavenworth, KS: School of Advanced Military Studies, U.S. Army Command and General Staff College, 1995; and Major Timothy Jones, *Attack Helicopter Operations in Urban Terrain,* Fort Leavenworth, KS: School of Advanced Military Studies, U.S. Army Command and General Staff College, December 1996. Steven Goligowski criticizes *FM 90-10* for being based on MOUT experience in World War II alone—see Major Steven P. Goligowski, *Future Combat in Urban Terrain: Is FM 90-10 Still Relevant?* Fort Leavenworth, KS: School of Advanced Military Studies, U.S. Army Command and General Staff College, 1995, and *Operational Art and Military Operations on Urbanized Terrain,* Fort Leavenworth, KS: School of Advanced Military Studies, U.S. Army Command and General Staff College, 1995. Also see Major Charles Preysler, *MOUT Art: Operational Planning Considerations for MOUT,* Fort Leavenworth, KS: School of Advanced Military Studies, U.S. Army Command and General Staff College, 1995.

[10]Outdated MOUT doctrine is a problem also recognized by the Russians, who use a MOUT doctrine based on their experience in The Great Patriotic War (World War II). See Colonel Mikhail Zakharchul, "View of a Problem," *Armeyskiy Sbornik,* translated by FBIS, FTS19970423002216, March 28, 1995.

are variously labeled low-intensity conflicts (LIC), operations other than war (OOTW), support and stability operations (SSOs), and small-scale contingencies (SSCs).

For example, research based on MOUT before 1982 implies that the number one factor explaining attacker "success" is isolation of the defender. But experience within the last ten years indicates that isolation of the defender may no longer be feasible. Rising urban populations and a decline in U.S. force structure make it increasingly difficult to isolate very large cities.[11]

Political considerations that restrict the use of combat force now complicate urban operations. World War II combat was usually high intensity, with little regard for noncombatant casualties, but today restrictive rules of engagement (ROE) often limit the application of power when noncombatants are present.

The objective of this monograph is to conduct an exploratory case study analysis to determine whether recent changes in the nature of urban operations are significant and what, if any, the policy implications are for U.S. Army doctrine. Three recent urban operations—Panama in 1989, Somalia in 1992–1993, and Chechnya 1994–1996—were chosen because they capture the possible range of political constraints.[12] This analysis concentrates on the key factors that determined the outcomes, with an emphasis on what is significantly "different" today. To determine how different, the lessons of the current three cases are compared to the lessons of 22 urban battles

---

[11]Russell Glenn highlights the Seoul example: in 1950, U.S. and South Korean forces recaptured Seoul when the Army's end strength was approximately equal to that city's population, about 1,000,000 people. Today the Army's strength is half that, while Seoul's population is about 13 million. See Russell Glenn, ". . . we band of brothers": The Call for Joint Urban Operations Doctrine, Santa Monica, CA: RAND, DB-270-JS/A, 1999, p. 12.

[12]Current Army doctrine arbitrarily defines three levels of MOUT that capture the level of political constraint: surgical MOUT, precision MOUT, and high-intensity MOUT. "Precision MOUT" is urban combat under *significantly* more restrictive ROE than high-intensity MOUT, and "surgical MOUT" includes operations that have a "surgical" nature, such as special-purpose raids, small strikes, and other small-scale combat actions. The scale of combat and the severity of political constraints differentiate between these levels.

fought between 1942 and 1982.[13]  The methodology of this work is the multiple-case study, an approach that will allow us to make analytic generalizations.[14]

For any research design, the author must know beforehand what questions to ask, what data are relevant to those questions, and how to analyze the results.  For the case study approach, there are five components of a research design that are especially important:[15]

1. The study's research questions

2. Its propositions

3. Its units of analysis

4. The logic linking the data to the propositions

5. The criteria for interpreting the findings

The primary research question is:  What are the dominant factors influencing success in recent urban warfare, and how do they compare to lessons learned from past urban battles?  In other words, have urban operations changed significantly?

The unit of analysis for each case study is defined to be a combat action that occurred in an urban environment.  For the cases analyzed here, the focus is on:

---

[13]In *Modern Experience in City Combat* (1987), R. D. McLaurin et al. distilled lessons from 22 urban conflicts ranging from Stalingrad to Beirut.  Because this report served as the basis for historical lessons that appeared in the official manual *Military Operations on Urbanized Terrain (MOUT)*, it can serve as a baseline with which to make comparisons.

[14]The multiple-case study should follow a replication logic, not a sampling logic.  The three case studies contained in this report are not meant to be a sample, and any conclusions that derive from them are not generalizable in a statistical sense.  What they do represent are three experiments that may be generalizable to a theoretical proposition.  These three cases can be thought of as three experiments in different settings.  The purpose of this monograph is to explore whether certain assumptions about the changing nature of urban operations are valid.  Sample size is irrelevant because a sampling logic is not used.  See Robert K. Yin, *Case Study Research*, Newbury Park, CA: Sage Publications, 1988, pp. 5–10.

[15]This section is based on the case study approach outlined in Yin, *Case Study Research*, op. cit.

- U.S. peace operations in Somalia, 1992–1993; specifically, the Mogadishu firefight of October 3–4, 1993.

- Operation Just Cause, 1989; specifically, the urban operations that occurred in cities, towns, airports, and bases throughout Panama in December.

- The Chechen War, 1994–1996; specifically, the battles for Grozny (December 1994–February 1995 and August 1996).

There are several reasons why these cases were chosen:

- All three cases are "modern" in that they all occurred within the last decade. These cases may offer historical lessons that are more relevant for U.S. urban operations doctrine in the post–Cold War environment.

- At least one side in each instance is a "conventional" force armed with modern weapons, either U.S. or Russian.

- A sufficient amount of literature and data exist to conduct a case study.

- They are similar to standard MOUT cases looked at in the past in that combat occurred and casualties were incurred on both sides; the essential unit of analysis—"the urban battle"—remains the same.

- Each case represents one of three classifications of MOUT as defined by U.S. doctrine: high intensity, precision, and surgical.[16] These classifications cover the spectrum of combat action within the MOUT environment (see Table 1).

Several supplemental research questions also motivate this work:

- For these three cases, are there common, significant factors that affected the outcome? Do these modern MOUT factors differ substantially from the significant factors found during earlier urban operation experiences? What variables are critical to attacker or defender success?

---

[16]Department of the Army, *An Infantryman's Guide to Combat in Built-up Areas,* Field Manual 90-10-1, p. G-1.

**Table 1**

**Three Classifications of MOUT**

| Case Study | Type of MOUT | Description |
| --- | --- | --- |
| Mogadishu firefight, October 3, 1993 | Surgical | MOUT is conducted by joint special operations forces; typical missions are special-purpose raids, small precision strikes, and small-scale personnel seizure and recovery operations. |
| Operation Just Cause, December 19–20, 1989 | Precision | MOUT is conducted by conventional forces to defeat an enemy intermixed with noncombatants. Strict ROE limit collateral damage and noncombatant casualties. |
| Battles for Grozny, December 1994–February 1995, August 1996 | High intensity | MOUT occurs over large built-up area such as an entire city; involves extensive destruction of the infrastructure, large conventional forces, less restrictive ROE. |

- How did the presence of larger numbers of noncombatants affect the course of operations?
- How did ROE affect the course of battle? Did they have a significant impact?
- Did any technologies or weapons make a profound difference?
- What role did aerospace play?
- Were combined arms, joint, or multinational teams crucial?
- What role did the media play?
- What were the critical nodes or centers of gravity?

The following theoretical propositions relate to the research questions above:

- Political, technological, and social developments in recent years have changed the way urban operations are justified and waged by democratic nations. Information technology, recent historical precedents, asymmetric responses to the growing conventional military dominance by the United States, changing standards of morality, and the shifting political justifications for the use of force have combined to exacerbate a longstanding geostrategic problem for conventional powers—how to wage restricted urban warfare while keeping casualties below some threshold of public tolerance.

- In the last decade, the following factors have all become relatively more significant: the presence of the media, the presence of noncombatants, ROE, information operation tools such as psychological operations (PSYOP), public affairs (PA), civil affairs (CA), and political-military strategy.

- Other important factors for urban operations—such as situational awareness, intelligence, airpower, surprise, technology, combined arms, and joint operations—have not changed in significance.

- Airpower has proved to be a mixed blessing for urban operations at the lower end of the intensity scale, especially because air-delivered munitions are still not discriminate enough to deal with the presence of noncombatants.

The remainder of the monograph is organized as follows. Chapter Two describes each case study in detail, starting with U.S. peace operations in Somalia (1992–1993), proceeding to Operation Just Cause (Panama, 1989), and finishing with the Chechen War (1994–1996). A cross-case analysis follows in Chapter Three, beginning with an outline of the lessons from past MOUT experience, based on other scholars' prior research of 22 urban battles between 1942 and 1982. An analytic framework is then presented to explain how broad changes over the last couple of decades may be driving significant changes in urban operations. Specific factors important to urban warfare are then addressed individually, especially with regard to how they affected the three very recent urban operations. The factors are noncombatants, ROE, the media, PSYOP and civil affairs, political-military strategy, situational awareness and intelligence, airpower, technology, surprise, combined arms, and joint operations.

# Chapter Two

# CASE STUDIES

Below is a brief narrative about each of the three historical cases. Each narrative includes some historical context, a description of the major events, the general strategy and tactics employed by both sides, and a quick summary of the dominant factors in the case. Readers already familiar with these cases may want to skip forward to Chapter Three, the cross-case analysis. A summary of the battle statistics for these three cases can be found at the end of this chapter.

## SURGICAL MOUT: Somali Peace Operations (1992–1993)

The peace operations that U.S. forces conducted in Somalia from 1992 to 1993 varied from humanitarian relief and assistance activities to peace enforcement operations. Combat occurred several times during this period, usually in urban areas and in the presence of noncombatants. The focus of this study is on the climatic battle of October 3–4, 1993. Only a few brief comments on the history leading up to the firefight are necessary.

## Setting the Stage

The United Nations Operation in Somalia (UNOSOM) was a massive international relief operation that ultimately sought to create a stable environment for the Somali people and address the underlying political and economic causes behind the famine devastating the country. UN Secretary-General Boutros-Ghali backed a UN resolution that created UNOSOM and expanded the UN's mission from a humanitarian relief operation to a nation-building operation that included

## Table 2

### Three Phases of Somali Peace Operations

| Operation | Dates |
|---|---|
| Provide Relief (UNOSOM I) | August 15 to December 9, 1992 |
| Restore Hope (UNITAF) | December 9, 1992 to May 4, 1993 |
| USFORSOM (UNOSOM II) | May 4, 1993 to March 31, 1994 |

disarming the population. The United States contributed a force known as the Quick Reaction Force (QRF) to the peace enforcement mission. Operation Restore Hope began on December 9, 1992, with an amphibious landing by U.S. Marines and some Navy SEALS on the beaches of Somalia. These were followed by additional units such as the U.S. Army's 10th Mountain Division.

The most powerful Somali warlord was Mohammed Aideed. In Aideed's view, the UN's goals amounted to a rejection of his claims to power in Somalia. His long-term strategy was to undermine and divide the coalition against him, keep the UN from reaching a negotiated settlement with other Somali clans, and force the strongest military power, the United States, to withdraw.[1]

Aideed used guerrilla tactics, including low-level attacks at weak targets, to avoid direct confrontations with UNOSOM forces. The Somalis were essentially urban guerrillas. They relied on stealth, surprise, dispersion, and concealment. The guerrillas operated without heavy logistical support, moved in small groups, and made do without heavy weapons. Their favorite offensive tactic was the tactical ambush. They avoided fixed fights and preferred to attack only when they possessed the advantage.

After the ambush of 24 Pakistani soldiers in Mogadishu on June 5, 1993, a UN Security Council resolution was issued to apprehend those responsible. On June 17, the UN arrest order was issued for the chief suspect, Aideed.[2] Aideed went into hiding and remained at

---

[1]Colonel William C. David, "The United States in Somalia: The Limits of Power," *Viewpoints*, 95-6, June 1995, located at http://www.pitt.edu.

[2]Rick Atkinson, "The Raid That Went Wrong; How an Elite U.S. Force Failed in Somalia," *Washington Post*, January 30, 1994.

large, continuing to attack UN peacekeepers with his Somali National Alliance (SNA) militia. On August 8 the SNA ambushed a U.S. military police convoy, killing four U.S. soldiers with a command-detonated mine. At this point, President Clinton ordered 130 Delta commandos, a Ranger company, and elements from the Army's special operations aviation unit to deploy to Somalia (otherwise known as Task Force Ranger). The search for Aideed was on, a search that would eventually include his lieutenants. The stage was set for what would turn out to be the most intense U.S. infantry firefight since the Vietnam War.

## Firefight of October 3–4

On the night of October 3, 1993, a company of U.S. Rangers and a Delta Force commando squadron fast-roped onto a gathering of Habr Gidr clan leaders in the heart of Mogadishu, Somalia.[3] The targets were two of Aideed's top lieutenants. The plan was to secure any hostages and transport them three miles back to base on a convoy of twelve vehicles. What was supposed to be a hostage snatch mission quickly turned into an eighteen-hour firefight when two Blackhawk helicopters crashed (see Figure 1). Eighteen Americans were killed in the fighting.

The helicopter assault force included about 75 Rangers and 40 Delta Force troops in 17 helicopters. The light infantry force on the ground was armed with small arms; the relieving convoys had nothing heavier than HMMWV-mounted .50 caliber machine guns and automatic grenade launchers. Close air support consisted of Blackhawk and Little Bird (AH-6) gunships. The Somalis were armed with assault rifles and rocket-propelled grenades (RPGs).

The Somalis knew that after the Rangers fast-roped in they would not be able to come back out on helicopters (the streets were too narrow). This meant a relief convoy would be necessary, so they immediately began setting up roadblocks all over the city.

---

[3]The picture of Mogadishu below and much of this section is drawn from the series of articles published in the *Philadelphia Inquirer* in November and December 1997 by Mark Bowden (the URL address is *http://home.phillynews.com/packages/somalia/sitemap.asp*) and his book, *Blackhawk Down: A Story of Modern War,* New York: Atlantic Monthly Press, 1999.

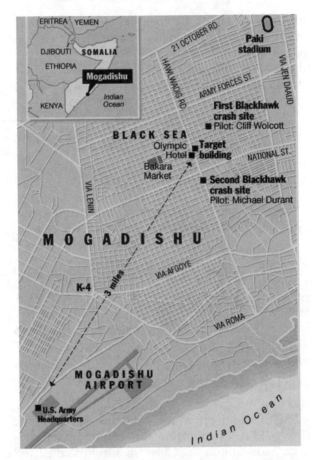

SOURCE: Mark Bowden, *"Blackhawk Down," The Philadel-phia Inquirer,* November 2, 1998, http://www.philly.com/packages/somalia/graphics/2nov16asp, and Matt Ericson, *The Philadelphia Inquirer,* staff artist. Reprinted with permission.

**Figure 1—Firefight:  October 3–4, 1993, Mogadishu**

The mission went well at first.  Twenty-four Somali prisoners were quickly seized at the target house.  Plans to haul them back to the airport base changed dramatically when a Blackhawk helicopter (Super 6-1) was shot down four blocks east of the target house.  Soon a second Blackhawk, Super 6-4, was shot down about a mile away.

An airmobile search and rescue force was sent to the Super 6-1 crash site and a light infantry force fast-roped down to secure the wounded crew. Task Force Ranger was also ordered to move to Super 6-1's crash site and extract the wounded crew. No rescue force was available to secure the second site, which was eventually overrun.[4]

The convoy holding the 24 Somali hostages was ordered to secure the second crash site but never made it. It wandered around the city suffering ambush after ambush until it eventually aborted the rescue attempt and returned to base. At one point, after about 45 minutes of meandering, this convoy ended up right back where it started. A second convoy of HMMWVs and three five-ton flatbed trucks was dispatched from the airport base to attempt a rescue at pilot Michael Durant's downed Super 6-4 Blackhawk, but those vehicles were also forced to turn back under heavy fire. At every intersection Somalis would open fire on any vehicle that came across.[5] Eventually a quick-reaction force of four Pakistani tanks, 28 Malaysian APCs, and elements of the 10th Mountain Division battled through barricades and ambushes to rescue Task Force Ranger at 1:55 A.M. on October 4.[6]

For the most part, U.S. commandos followed standard doctrine for city fighting. Using fire and maneuver, teams and squads leapfrogged each other. At times, infantry moved out on foot to cover the convoy from both sides of the street. The main problem was that the convoys kept halting, exposing vehicles located in the middle of street intersections to concentrated enemy fire.

There was a Somali battle plan of sorts. Aideed's SNA militia (between 1,000 and 12,000 men) was organized to defend 18 military sectors throughout Mogadishu. Each sector had a duty officer on

---

[4]According to Mark Bowden in *Blackhawk Down*, one of the flaws in the mission planning was this lack of a second rescue force. Nobody had taken seriously the prospect of two helicopters going down.

[5]Fortunately for the Americans, the ambushes were poorly executed. The correct way to ambush is to let the lead vehicle pass and suck in the whole column, then open fire on the unarmored flatbed trucks in the middle. The Somalis usually opened up on the lead vehicle. They also cared little for fratricide. Because Somalis fired from both sides of the street, they certainly sustained friendly fire casualties.

[6]Rick Atkinson, "Night of a Thousand Casualties; Battle Triggered U.S. Decision to Withdraw from Somalia," *Washington Post*, January 31, 1994, p. A11.

alert, connected through a crude radio network.[7] By the time the U.S. assault team had landed, the Somalis were burning tires to summon all militia groups.

The most likely tactical commander of the Somalis during the October 3–4 fight was Colonel Sharif Hassen Giumale, who was familiar with guerrilla insurgency tactics. Guimale's strategy was to fight the Americans by using barrage RPG fire against the support helicopters, ambushes to isolate pockets of Americans, and large numbers of SNA militiamen to swarm the defenders with sheer numbers.

Somali tactics were to swarm toward the helicopter crashes or the sound of firefights. Out in the streets, militiamen with megaphones shouted, "*Kasoobaxa guryaha oo iska celsa cadowga!*" (Come out and defend your homes!). Neighborhood militia units, organized to stop looters or fight against other enemy clans, were united in their hatred of the Americans. When the first helicopter crashed, militia units from the surrounding area converged on the crash sites along with a mob of civilians and looters. Autonomous militia squads blended in with the masses, concealing their weapons while they converged on the Americans.

Most of the Somalis were not experienced fighters. Their tactics were primitive. Generally, gunmen ducked behind cars and buildings and jumped out to spray bullets toward the Rangers. Whenever Americans moved, the Somalis opened up from everywhere. Gunmen popped up in windows, in doorways, and around corners, spraying bursts of automatic fire.

From a military viewpoint, the October battle in Mogadishu was a tactical defeat for the Somalis—the Ranger and Delta commandos were able to complete their mission and extract the hostages. In terms of relative casualties, the mission was also an American military success—only eighteen American soldiers were killed and 73 wounded while more than 500 Somalis died and at least a thousand were put in the hospital.[8] But from a U.S. strategic or political view-

---

[7]Atkinson, "The Raid That Went Wrong."

[8]Atkinson reports the same number of Americans killed but 84 wounded. He also reported 312 Somali dead and 814 wounded. See Atkinson, "Night of a Thousand Casualties," p. A11.

point, the battle not was a success because the end result was an American withdrawal from Somalia. On November 19, 1993, President Clinton announced the immediate withdrawal of Task Force Ranger, and he pledged to have all U.S. troops out of Somalia by March 31, 1994. The casualties incurred were simply too high a cost for the U.S. national interests at stake in Somalia.

The strategic ramifications of this battle persist. The U.S. decision to withdraw from Somalia after losing relatively few soldiers has had unintended consequences—many adversaries now question American resolve and its obsession with casualties. In a May 28, 1998, ABC news interview, the terrorist Osama bin Laden echoed this sentiment:

> We have seen in the last decade the decline of the American government and the weakness of the American soldier who is ready to wage cold wars and unprepared to fight long wars. This was proven in Beirut when the Marines fled after two explosions. It also proves they can run in less than 24 hours, and this was also repeated in Somalia.[9]

Some people believe that the low casualties of the Persian Gulf War have lulled the American public into unrealistic expectations about the price of modern combat. Others argue that the U.S. public has never been willing to tolerate casualties when national interests are not at stake.[10] Regardless of which argument is correct, the important point to realize is that if the enemies of the United states believe the American people do not have the "stomach" to take casualties, they will act in accordance with this belief.

---

[9]G. E. Willis, "Remembering Mogadishu: Five Years After the Firefight in Somalia, Some Say U.S. Forces Abroad Still Are Reeling from It," *Army Times*, October 1998, p. 16.

[10]One study found that the unwillingness of the public to tolerate very high casualties in Lebanon and Somalia had to do with the fact that majorities—and their leaders—did not perceive the national interests at stake important enough to justify much loss of life. In addition, the absence of foreign policy consensus among leaders will make the public more sensitive. See Eric Larson, *Casualties and Consensus: The Historical Role of Casualties in Domestic Support for U.S. Military Operations,* Santa Monica, CA: RAND, MR-726-RC, 1996, p. xvi.

## Dominant Factors

Aideed's victory was due to several factors. The nature of the urban terrain had an inhibiting effect on U.S. situational awareness and firepower. The support of the indigenous population for their militia helped to conceal insurgents and hinder the use of airpower. Somali RPGs changed the whole course of the mission when two U.S. Blackhawks were downed. The absence of heavier U.S. armor and lack of combined arms were sorely felt, especially when roadblocks needed to be cleared. Finally, the Somalis were willing to take casualties and could afford to follow their costly swarm tactics.

## PRECISION MOUT: Operation Just Cause, Panama (1989)

During Operation Just Cause (OJC) American joint forces attacked the Panamanian Defense Force (PDF) using strict ROE. Combat actions included airfield seizures and deliberate attacks on fortified positions. Urban targets were positioned among the cities, airports, military bases, and rural areas.

## Setting the Stage

The United States has a long history of intervention in Panama, from the first Marine landing in 1903 to the permanent stationing of U.S. troops to protect the Panama canal. In the 1980s, Manuel Antonio Noriega rose to power and became a useful asset to the United States for his contributions to intelligence and counternarcotics operations. Later allegations that Noriega had rigged the 1984 election eventually led to the suspension of U.S. military and economic aid. After the Iran-Contra scandal broke, Noriega's usefulness continued to decline, especially when he continued to curtail constitutional rights in Panama. Tensions escalated as he stepped up harassment of U.S. military personnel and tried to stoke anti-American sentiment among his own people. In February 1988, two Florida grand juries indicted Noriega on separate charges related to his drug cartel connections. The Reagan administration applied financial pressure on Panama, invoking formal sanctions in April 1988. As a war of words continued between Noriega and the Bush administration, the PDF continued its hostile behavior toward U.S. servicemen.

The trigger event for Operation Just Cause (OJC) was the killing of a U.S. officer on December 16, 1989. U.S. forces were ordered to overthrow the Noriega dictatorship, create a safe environment for U.S. citizens living in Panama, secure the Panama Canal, and reinstate a democratically elected government.

## The Assaults on Panama and the Aftermath

On December 19, 1989, units from the Army, Navy, and Air Force assaulted 27 critical objectives throughout Panama, the largest airborne operation since World War II. Initial targets included PDF concentrations, garrisons, and airports, as well as media, transportation, and command and control nodes. Joint Task Force South conducted the attack with the 13,000 U.S. troops already garrisoned in Panama and another 13,000 deployed troops from the United States.[11] Most of the fighting and many of the most crucial assignments went to special operations forces.[12] The opposing force, the PDF, contained about 15,000 men with an effective combat strength of about 6,000.[13] The heaviest PDF threat was 28 armored cars.

The U.S. commander, Lt. General Carl Stiner, hoped to paralyze the PDF by hitting every vital node with overwhelming force.[14] The simultaneous assault on dozens of targets in the middle of the night proved effective against the highly centralized PDF. The major targets were the locations of PDF reinforcements, two airfields, a few

---

[11]The several units already stationed in Panama included a battalion of the 7th Special Forces Group at Fort Davis; the Jungle Operations Training Center at Fort Sherman, which could field a battalion of troops; and USARSO-controlled troops from the 7th Infantry Division (Light) and 5th Infantry Division (Mechanized).

[12]Although the elite forces totaled only 4,150 troops compared to the remaining 23,000 regular American troops, they took the brunt of the losses.

[13]Malcolm McConnell, *Just Cause: The Real Story of America's High-Tech Invasion of Panama,* New York: St. Martin's Press, 1991, p. 30. Other sources indicated that PDF combat strength was lower, around 4,000 troops. See Ronald M. Cole, *Operation Just Cause: The Planning and Execution of Joint Operations in Panama, February 1988–January 1990,* Washington, D.C.: Office of the Chairman of the Joint Chiefs of Staff, 1995, p. 37.

[14]Stiner was also commander of the XVIII Airborne Corps. His boss was General Maxwell Thurman, Commander in Chief, U.S. Southern Command (CINCSO). Joint Task Force South (JTFSO) was set up to execute the Blue Spoon operations order, the plan for the entire operation.

bridges, a naval base, and the main PDF stronghold in Panama City, *La Comandancia*.[15]    U.S. special forces also attempted to snatch Noriega himself.[16]

The most difficult PDF urban operations target proved to be *La Comandancia* in Panama City, a compound of fifteen buildings surrounded by a ten-foot wall right in the middle of the city. It was the command and control center of Noriega's forces as well as an armory and motor pool. U.S. mechanized infantry paved the way for light infantry movement toward the PDF strongholds in the heart of the city. APCs and dismounted infantry gradually contracted a circle around the compound, seizing key intersections overlooking the stronghold and clearing snipers from the vicinity. Under the cover of supporting air strikes and Sheridan tank fire from supporting positions on a nearby hill, dismounted troops were eventually able to blast a hole through the wall of *La Comandancia* with demolitions.

The PDF did a poor job utilizing their stockpile of RPGs. The urban terrain around the compound offered numerous opportunities for ambushing the relatively light American vehicles that were covering the approaches, yet the PDF only managed to take out one M113 armored personnel carrier and temporarily halt two columns at roadblocks.

The Ranger assault on the Rio Hato military base was one of the biggest firefights of OJC. Two battalions of Rangers parachuted into the Panamanian military base, located about 75 miles west of Panama City. The Ranger light infantry was supported by a pair of new Apaches, a Spectre gunship, AH-6 "Little Bird" helicopters, and Stealth F-117As. The fighting in the barracks area was classic MOUT—building to building, room to room. The PDF fought stubbornly, retreating out the rear of buildings to ambush the pursuing

---

[15]The PDF was primarily a ground force organized into thirteen military zones totaling two battalions, ten additional infantry companies, and a special forces command. The heaviest equipment it possessed was armored cars (V-300). Some paramilitary forces were available also. PDF naval forces consisted of 13 vessels, including fast patrol boats, and the PDF air force had 38 fixed-wing aircraft, 17 helicopters, and some air defense guns. See Thomas Donnelly, Margaret Roth, and Caleb Baker, *Operation Just Cause: The Storming of Panama*, New York: Lexington Books, 1991, p. 75.

[16]They tried more than 40 times, failing every time. See Donnelly, Roth, and Baker, *Operation Just Cause*, p. 105.

Rangers from gullies and other cover. In this action, the Rangers lost 4 dead and 18 wounded (another 23 had been injured in the jump), but they killed 34 PDF soldiers, captured 362, and detained 43 civilians.[17]

OJC could easily have turned into a nightmare for U.S. planners. Noncombatant casualties, especially American civilians, were a major concern. Many Americans lived, worked, or went to school right next to Panamanians. One of the task forces involved in the operation, Task Force Atlantic, was solely responsible for the safety of a thousand Americans living on joint U.S.-Panamanian military installations or in civilian housing.[18]

For Operation Just Cause as a whole, 23 American soldiers and 3 American civilians were killed, and 324 were wounded. At least 314 PDF soldiers were killed in the fighting, and between 200 and 300 Panamanian civilians perished.[19]

## Dominant Factors

OJC was a decisive American victory for many reasons. The fact that U.S. forces were already stationed in Panama and had been training there for years granted enormous advantage. The U.S. operation used operational maneuver, mainly through airlift, to finish the fight throughout Panama in just a few days. The forces were able to do this because no surface-to-air missile (SAM) threat was present.

The PDF was generally of poor quality, with most of its soldiers quite unwilling to fight to the death. PDF troops usually began to desert once the fighting began, as they did at Cimmarron. The PDF was also caught by surprise.

Finally, the indigenous population was not united behind Noriega's oppressive dictatorship but was split in its loyalty. As one soldier put it: "There [were] people out partying and waving U.S. flags and cheering for us. And then we would turn a corner and start heading

---

[17]Ibid., p. 349.

[18]Ibid., p. 237.

[19]Ibid., p. 390.

down another way, and all of a sudden we'd start getting shot at."[20] This unpredictability complicated MOUT for both sides. The PDF's only real chance to win was to somehow protract the conflict and inflict unacceptable casualties on the United States, forcing a domestic political response that would end the fighting. Without the support of the population, that was impossible.

## HIGH-INTENSITY MOUT: The War in Chechnya (1994–1996)

The Chechen War (1994–1996) ran the gamut of urban operations—from the surgical strikes of Budyonnovsk and Kizlyar-Pervomaiskoye to high-intensity MOUT within the city of Grozny. Chechnya is also the most recent example of how an insurgent force defeated a conventional military power by means of a superior political-military strategy. The irony in this case is that a relatively small force of insurgents defeated an army that arguably has the most MOUT experience of any force in the world.[21]

### Setting the Stage

Chechnya has been fighting Russian domination for over 250 years. Tsarist, Bolshevik, and Soviet forces have all put down Chechen revolts. From the Stalinist purges of the 1920s and 1930s to the mass deportation of the Chechen people to Siberia in 1944, the Chechens have accumulated many reasons for hating Russians. In 1991, after the August coup in the former Soviet Union, nationalist leaders in the Republic of Chechnya saw an opportunity to press demands for Chechen independence. Soon after, President Dzhokar Dudayev formally declared Chechen independence.

Chechnya was geostrategically important to Russia for many reasons. Conflict raged across the region (between the Azerbaijans and Armenians, between the Ingush and the North Ossetians, and between Georgia government forces and an Abkhazian separatist movement). Major Russian oil pipelines ran from the Caspian basin through

---

[20]Quote from Sergeant Joseph Ruzic in Donnelly, Roth, and Baker, *Operation Just Cause*, p. 313.

[21]Not even counting World War II experience, the Russians conducted MOUT in Berlin (1953), Budapest (1956), Prague (1968), and Kabul (1979).

Chechnya and the Transcaucasus to the Black Sea. The Russians were also concerned that if the Republic of Chechnya were allowed to break free of the Russian federation, other minority republics in the North Caucasus might seize upon the precedent to demand their own independence.[22]

After President Dudayev dissolved the Chechen parliament in 1993, an opposition group developed and small-scale violence erupted between contending parties. Dudayev refused to negotiate a return to the Russian federation, and after a covert Russian attempt to support Dudayev's political opposition was thwarted and exposed, Boris Yeltsin decided to send a peacemaking force into Chechnya on December 11, 1994.

A Russian force of about 23,800 men, 80 tanks, 208 APC/IFVs, and 182 artillery pieces invaded Chechnya.[23] The Russians advanced into Chechnya along three axes of advance—one each from the north, east, and west—in order to isolate and attack the capital city of Grozny (see Figure 2). Before Grozny was encircled or blockaded, however, the western force of 6,000 Russian soldiers mounted a mechanized attack on New Year's Eve 1994.

The Chechens started the war with about 35 tanks, 40 armored infantry vehicles, 109 artillery pieces, multiple rocket launchers, mortars, and air defense weapons.[24] They also had a large arsenal of

---

[22]See Timothy Thomas, *The Caucasus Conflict and Russian Security: The Russian Armed Forces Confront Chechnya, Part I and II,* Fort Leavenworth, KS: U.S. Army, Foreign Military Studies Office, 1995, p. 4.

[23]Later, the Russian force grew to 38,000 men, 230 tanks, 454 APC/IFVs, and 388 artillery pieces. Lieven believes that about 40,000 Russian troops entered Chechnya, but because many Russian units were seriously understrength, the number may have been as low as 20,000. See Anatol Lieven, *Chechnya: Tombstone of Russian Power,* New Haven and London: Yale University Press, 1998, p. 280.

[24]The figures quoted are from "Russia's War in Chechnya: Urban Warfare Lessons Learned 1994–96," *Marine Corps Intelligence Activity Note,* CBRS Support Directorate (MCIA-1575-xxx-99), November 1998, p. 4. Another source indicates that the Chechens had 40–50 T-62 and T-72 tanks, 620–650 grenade launchers, 20–25 "grad" multiple rocket launchers, 30–35 APCs and scout vehicles, and 40–50 BMPs. See Surozhtsev, "Legendary Army in Grozny," *Novoye Vremya,* No. 2–3, January 1995, pp. 14–15.

RPGs.[25] Dudayev's forces in Grozny probably numbered about 15,000 men.[26]

## The First Battle for Grozny (December 1994–January 1995)

The Russian plan was to seize important Chechen nodes such as the presidential palace, railroad station, government, and radio and television buildings. The main attack focused on the railway station, several blocks southeast of the palace.

Disregarding proper combined arms tactics, Russian armored vehicles drove into Grozny without deploying dismounted infantry support, allowing Chechen infantry to ambush the tanks in the spearhead. In the 131st Motorized Brigade, only 18 out of 120 vehicles escaped destruction.[27] Without infantry, Russian tanks were easy pickings for the waiting Chechens armed with RPGs:

> The Russians stayed in their armor, so we just stood on the balconies and dropped grenades on to their vehicles as they drove by underneath. The Russians are cowards. They can't bear to come out of shelter and fight us man-to-man. They know they are no match for us. That is why we beat them and will always beat them.[28]

---

[25]Chechen anti-tank weapons included Molotov cocktails, RPG-7s (including -7B, -7B1, -7D variants), RPG-18s, and long-range systems such as the Fagot (24 systems), Metis (51 systems), and 9M113 Konkurs antitank (2 systems). They also had the PG-7VR system for reactive armor targets. See "Russian Military Assesses Errors of Chechnya Campaign," *International Defense Review*, Vol. 28, Issue 4, April 1, 1995; and Aleksandr Kostyuchenko, "Grozny's Lessons," *Armeyskiy Sbornik*, translated in FBIS FTS19951101000633, November 1, 1995.

[26]According to Andrei Raevsky. Raevsky also cites Russian sources indicating 10,000 Chechens were waiting in Grozny. See Andrei Raevsky, "Russian Military Performance in Chechnya: An Initial Evaluation," *Journal of Slavic Military Studies*, Vol. 8, No. 4, London: Frank Cass, December 1995. Thomas cites Russian estimates of 11,000–12,000 Chechens. See Thomas, *The Caucasus Conflict and Russian Security*, p. 30.

[27]Baseyev's claim that 216 Russian vehicles were destroyed is probably exaggerated. General Pulikovskiy says only 16 were hit. See Mikhail Serdyukov, "General Pulikovskiy: Fed Up!" *Sobesednik*, translated in FBIS, September 1996.

[28]See Lieven, *Chechnya: Tombstone of Russian Power*, p. 109.

Figure 2—Russian Axes of Advance into Chechnya (1994)

After the devastating losses of January 1–3, the Russians adjusted their tactics.[29]  They learned the hard way that tanks should be well protected by screening infantry and should be used for fire support just outside of RPG range.  They pulled back from the center and pounded the city with artillery and airpower.  ROE were discarded.

---

[29]Ninety percent of Russian losses in the assault on Grozny occurred in the first few days between December 31, 1994, and January 2, 1995.

Special shock troops, paratroopers, motorized infantry units, and marines systematically pushed the Chechens back building by building. This initial battle for Grozny lasted several weeks.

Combat operations broke down into small unit firefights because of the nonlinear nature of urban terrain. Commanders sometimes could not exercise command and control over adjacent units because of a lack of common corridors or entrances. If a Russian unit advanced too far (or adjacent units fell back), it was cut off, surrounded, and attacked by Chechens, like "wasps on a ripe pear."[30]

By January 10 the Russians had managed to make two corridors into the city to resupply and evacuate the wounded. Dudayev's forces fought back fiercely, especially in the center of the city. A cease-fire began on the 10th, was violated by both sides, and officially ended on the 12th. Using fresh reinforcements, the Russians renewed the offensive, pounding Chechen positions in the city center with artillery.[31] As the Russian assault continued on the 13th and 14th, MVD forces blocked the main departure routes out of Grozny, effectively sealing it off by the 15th. After two Russian bombs penetrated to the basement of the palace, the Chechen on-scene commander, Maskhadov, retreated to the southeastern part of Grozny to prepare for a general evacuation to the mountains.[32] The Russians gained the palace on the 19th. Somewhat demoralized by the symbolic loss of the palace, many Chechen rebels began leaving the city, moving in southerly and southeastern directions.

Through most of January, the Russians failed to encircle Grozny and the Chechens continued to resupply their forces. Spurred to fight for their homes and families, Chechen volunteers flowed into the city from the countryside.[33] Small groups of Chechens continued to seek

---

[30]See Serdyukov, "General Pulikovskiy: Fed Up!"

[31]At one point a round landed every 10 seconds for over three hours. See Timothy Thomas, "The Caucasus Conflict and Russian Security: The Russian Armed Forces Confront Chechnya, III. The Battle for Grozny, 1–26 January 1995," *Journal of Slavic Military Studies,* London: Frank Cass, Vol. 10, No. 1 (March 1997), p. 75.

[32]Ibid., p. 76.

[33]Although the Chechen "high command" did not exercise control over much of the Chechen resistance, it did issue calls for volunteers to stream into the cities like Grozny on their own and jump into the fight.

and ambush small Russian units in the porous urban terrain. A pattern set in: the Chechens would hide in basements during the daylight barrages, then emerge for hit and run attacks at night.

It was not until the 21st that Russian task forces Group West and Group East fought their way to the center of Grozny, at which point they basically controlled about half the city. Grozny was finally cleared of rebels around late February.

## Russian Strategy and Tactics

Since the Chechen War evolved over several weeks of combat and was far larger in scope than Operation Just Cause and the Mogadishu firefight, the development of strategy and tactics deserves a special mention.

The Russian strategy that evolved was to bully the cooperation of the people in order to cut off support for the Chechen fighters. Towns and villages were pounded from the air until they signed individual truces with Russian forces.

Their poor tactics in the first assault on Grozny notwithstanding, the Russians had a well-developed doctrine for urban warfare based on their extensive experience both before and after World War II.[34] The problem was that urban operation skills were a lost art among most active duty soldiers because MOUT training was almost nonexistent. Eventually they did manage to relearn the tactics, techniques, and procedures (TTPs) involved in isolating a city, establishing a foothold, and clearing the city sector by sector. They used direct-fire artillery, RPGs, automatic grenade fire, and machine guns to provide suppressive fire, smoke bombs to cover approaches to building objectives, demolitions to create entryways, and small teams of infantry to clear buildings room by room. Special assault units proved to be the most effective fighting formations.[35]

---

[34]The Russians had also developed a counterinsurgency doctrine in Afghanistan to fight a nonlinear battle with raids and ambushes using spetsnaz, airborne, and air assault units. They found that the key to nonlinear counterinsurgency operations was decentralized command and independent brigade and battalion operations.

[35]The Russians basically reinvented the wheel—the lessons they learned in World War II—by creating special units consisting of three mechanized infantry platoons, two

## Chechen Strategy and Tactics

For the Chechens an outright military victory was unlikely, so their goal was to inflict as many casualties as possible on the Russian people and erode their will to fight. The Chechens used an "asymmetric" strategy that avoided battle in the open against Russian armor, artillery, and airpower. They sought to even the fight by fighting an infantry war. Time and again, the Chechens forced their Russian counterparts to meet them on the urban battlefield where a Russian infantryman could die just as easily as a Chechen fighter.

The Chechen strategy has been described as the battle for "successive cities."[36] After Grozny fell, the Chechens moved their operations base to Argun, Shali, and other cities to continue the battle of urban attrition. Dudayev deliberately used cities throughout Chechnya as strategic strongpoints from which to defend his country.[37] As one Chechen put it, "We were very happy they came into the city because we cannot fight them in an open field."[38]

Overall, the Chechens used a mobile area defense. A fixed defense based on strongpoints was vulnerable to Russian firepower, so the

---

flame-thrower platoons (each with nine Shmel launchers), two air defense guns, one minefield-breaching vehicle, a combat engineer squad, a medical team, and a technical support squad. The minefield-breaching vehicle was the UR-77, which used a rocket-propelled line charge launcher mounted on the rear hull for explosive breaching of mine fields. The Shmel flame-thrower was a favorite among Russian troops. Called "pocket artillery," the Shmel is a single-shot, disposable weapon that looks like a U.S. light antitank weapon (LAW). It was used against places with confined spaces—such as bunkers or interior rooms—and performed like a fuel air explosive. It was also an effective anti-sniper weapon. See "Russia's War in Chechnya: Urban Warfare Lessons Learned 1994–96," p. 9.

[36]See Timothy Thomas, unpublished draft manuscript, "Some Asymmetric Lessons of Urban Combat: The Battle of Grozny (1–20 January 1995)," Fort Leavenworth, KS: U.S. Army, Foreign Military Studies Office, January 8, 1999.

[37]Baev argues that this strategy worked to Russian advantage because once the weather improved and air-to-ground coordination improved among Russian forces, pinpoint air strikes enabled the rapid destruction of Dudayev's strongholds at Argun, Gudermes, and Shali. See Pavel K. Baev, "Russia's Airpower in the Chechen War: Denial, Punishment and Defeat," *Journal of Slavic Military Studies*, Vol. 10, No. 2, London: Frank Cass, June 1997.

[38]See Michael Spector, "Commuting Warriors in Chechnya," *The New York Times*, February 1, 1995.

Chechens relied more on a fluid and elusive hit-and-run defense.[39] The mobile Chechens used back alleys, sewers, basements, and destroyed buildings to slip around and through Russian lines. Chechen vehicle detachments transported supplies, weapons, and personnel quickly and easily throughout Grozny.  Chechen artillery deployed near schools or hospitals, fired a few rounds, and dispersed to avoid counterbattery fire.[40]

At the tactical level, the loose organization and command of most of the Chechen volunteer force had both positive and negative aspects. On the one hand, independent groups of autonomous units could operate efficiently in the fluid, nonlinear, urban battlefield, helping to alleviate the complex command and control problem.  On the other hand, a lack of discipline and responsibility to higher command led some groups to abandon their posts when they got bored or heard shooting elsewhere, leaving crucial posts undefended.[41] Most bands wandered about without overall command coordination.

Extensive use of the ambush, fighting at night, and the use of anti-tank hunter-killer teams were the hallmarks of Chechen tactics.[42] Roving bands of 10–15 men (who could further subdivide into 3- to 4-man cells) would swarm toward the sound of Russian engines and volley fire RPG-7 and RPG-18 antitank missiles from upper-floor windows.[43]  Chechens used classic ambush techniques:  wait for a column of vehicles to wander all the way into a kill zone, take out the

---

[39]This is not to say that strongpoints were ignored.  Three defensive belts were constructed in Grozny.  The inner belt consisted of five major fortifications across the streets leading to the presidential palace.  See Carl Van Dyke, "Kabul to Grozny: A Critique of Soviet Counter-Insurgency Doctrine," *Journal of Slavic Military Studies*, Vol. 9, No. 4, London:  Frank Cass, December 1996, p. 698.

[40]Thomas, "Some Asymmetric Lessons of Urban Combat."

[41]For example, soldiers held no rank.

[42]It should be noted that most Chechens were for the most part inexperienced, although some had fought in Afghanistan, in the Nagorno-Karabakh regional conflict between Azerbaijan and Armenia, or in the Abkhazia region of Georgia.  Chechens were generally excellent shots, most having learned to use a rifle at an early age.  The most successful Chechen RPG gunners were usually teenagers as young as 16.  Based on commends by Arthur Speyer, RAND/TRADOC/MCWL/OSD Urban Operations Conference, Santa Monica, California, March 22, 2000.

[43]One Chechen described battle group size as 20 to 50 people.  See Spector, "Commuting Warriors in Chechnya."

leading and trailing vehicles to create a trap, and finish off the rest of the vehicles one by one, shooting any survivors as they bailed out. Russian tank armor proved vulnerable to top attack.[44] The Chechens also booby-trapped bodies, buildings, and obstacles—anything that Russian soldiers might have to move or clean up.[45]

## The Rest of the Chechen War, Including the Second Battle for Grozny in August 1996

After the first battle for Grozny, Chechen forces continued their retreat southeast to the cities of Gudermes and Shali. When the Russians moved to encircle these smaller cities with armor, the Chechens were forced to evacuate to avoid being captured. Taking casualties from heavy artillery and air bombardment, the Chechens quickly withdrew from villages in the flatlands to the forested foothills and mountains where it was impossible for Russian tanks and IFVs to operate. By May 1995, the Russians controlled the main cities and towns and the Chechens were forced to hole up in the mountains (see Figure 2).[46]

Russian commanders thought the Chechens would find little cover in the mountains, becoming more vulnerable to their airpower.[47] They also believed the Chechens would find it difficult to support a modern partisan army without aid from abroad. However, the Chechens managed to force a temporary cease-fire with their successful surgi-

---

[44]Ninety-eight percent of destroyed Russian tanks were hit where reactive armor could not be placed. The top armor of the T-72 and T-80 tanks, especially the turret roof and engine deck, was thin and easily penetrated by the shaped charge of an RPG warhead. The Russian vehicles were also hampered by an inability to elevate their crew-served weapons in defense.

[45]Thomas refers to an FBIS article that describes how the Chechens sometimes would put a pet inside a booby-trapped building to attract attention. See Thomas, "The Caucasus Conflict and Russian Security: The Russian Armed Forces Confront Chechnya, III. The Battle for Grozny, 1–26 January 1995," p. 82.

[46]Geographically, Chechnya lies on the north side of the Caucasus mountain range, more than 1,000 miles south of Moscow. The northern part is a grassy plain, but the heartland is in the south, a wild and rugged region where rivers thread their way through gorges from the ridge of the Caucasus. The high hills are still covered with thick beech forests, useful terrain for guerrilla operations.

[47]See Steve Erlanger, "Russian Troops Take Last Chechen Cities," *The New York Times*, April 1, 1995.

cal attack on the Russian city of Budyonnovsk, giving them time to reorganize and consolidate.

The turning point of the war came on August 1996, when the Chechens launched a surprise counteroffensive on Grozny, Argun, and Gudermes to demonstrate to the Russian people that the insurgents could still strike when they wanted. As the North Vietnamese did in their Tet offensive in 1968 in South Vietnam, the Chechens launched a costly attack that no doubt would eventually have ended in failure to achieve a strategic and political goal, in this case to embarrass Yeltsin. The Russian president had just proclaimed the war over and was getting ready to celebrate his inauguration for a second term.[48] The second battle for Grozny made it obvious to the Russian people that the war was far from over.

Over 1,500 Chechen fighters infiltrated on foot into the city to attack Russian army posts, police stations, and key districts. The entire 12,000-man Russian garrison was pinned down under mortar, machine gun, and sniper attack. Poor Russian morale and a lack of necessary troop strength allowed the Chechens to infiltrate into Grozny with impunity.[49] Over the course of the next several days, Russian relief columns, tanks, and IFVs attempted to relieve their besieged outposts. All the Russian columns were ambushed and destroyed.

During the Chechen counterattack on Grozny, the Russians lost 500 dead, 1,407 wounded, 182 missing, and an unknown number of casualties among the 300,000 civilians present. Political will power for the war evaporated.[50] By the end of August, Russian national security adviser Alexander Lebed had brokered a peace deal with Chechen commander Aslan Maskhadov that avoided declaring a victory for either side. It was plain to all who the victor was when all Russian forces were ordered to evacuate Grozny.

---

[48]See Carlotta Gall and Thomas de Waal, *Chechnya: Calamity in the Caucasus,* London and New York : New York University Press, 1998, p. 332.

[49]Later, a second wave of 1,500 reserve Chechen fighters infiltrated the city.

[50]Other factors helped ease the acceptance of a peace agreement. By this time Dudayev was dead. The Russians found his replacement, Maskhadov, much easier to work with. One of the original political reasons for invading Chechnya—the fear of a Caucasus chain reaction of exodus from the Russian Federation—had proved unfounded.

The Chechen War lasted twenty months, killed some 50,000 civilians, 6,000 Russian soldiers, and 2,000–3,000 Chechen fighters, and resulted in an agreement to put off the question of Chechen independence for five years. The Chechens were able to assert their independence from Moscow and Yeltsin was forced to remove all Russian forces from Chechnya.

## Dominant Factors

The Russians paid heavily for their attacks on the cities of Chechnya for many reasons, most of them related to the steady erosion of the Russian military since the end of the Cold War.[51] Given the number of problems, it would be tedious to list every possible factor that might have influenced the outcome. There were many problems: poor command and control, a shortage of troops, poor training, the refusal of units and commanders to execute orders, low morale, and poor logistics are but a few. This analysis merely describes the significant factors that determined the outcome of this war.[52]

Poor tactics was certainly the main reason for excessive early losses. Sending Russian armor straight into Grozny without infantry support allowed the Chechens to ambush Russian vehicles from overlooking buildings and street corners. The Russians also suffered from poor unity of command at all levels, highlighted by the absurd example of Yeltsin's declaration of a cease-fire while Russian military commanders simultaneously launched offensive attacks.[53]

---

[51]In particular, the budget cutbacks after the collapse of the former Soviet Union severely lowered training, morale, and troop quality.

[52]The focus of the cross-case analysis is on the factors related to recent changes in the nature of MOUT.

[53]Raymond Finch argues that poor leadership was the main reason why the Russians failed. The issue of absurd orders, the casual disregard for the fate of soldiers, the abysmal conditions of the common soldier, and general corruption were the main leadership failures. See Major Raymond C. Finch III, *Why the Russian Military Failed in Chechnya*, Fort Leavenworth, KS: U.S. Army, Foreign Military Studies Office, 1998. Effective joint operations were difficult, considering the number of services involved. In addition to Ministry of Defense (MoD) forces (generally referred to as the "army"), there were Border troops, Interior troops, the Presidential guard, and forces belonging to 13 other ministries. See Charles J. Dick, "A Bear Without Claws: The Russian Army in the 1990s," *Journal of Slavic Military Studies*, Vol. 10, No. 1, London: Frank Cass, March 1997.

Lack of training was another important factor behind the Russian failure. Even basic individual and unit skills essential to any combat environment were seriously underdeveloped because of the catastrophic budget cuts of the early 1990s.[54] Some servicemen did not know how to dig a foxhole, lay mines, prepare sandbagged positions, or fire a machine gun, let alone conduct urban operations.[55] The Russian army had not had a division-level field exercise since 1992. Helicopter pilots had less than a third of their required flight training. Russian deficiencies in urban operations tactics and training led to appalling losses.

The lack of a professional noncommissioned officer corps in the Russian army only exacerbated the training problem.[56] Planning suffered as a result.[57] In 1994, Russian units lacked sufficient numbers of small-unit leaders such as platoon and squad leaders, positions that are crucial in urban operations.[58]

The Chechens won for many reasons, not the least of which was a defensive strategy that utilized urban operations to negate Russia's firepower advantage. The Chechens enjoyed some crucial advantages: they fought on their own turf, spoke Russian, and in many cases had served in the Russian army. In the initial battle for Grozny,

---

[54]Declining recruit quality also exacerbated the training problem. Morale and discipline had sagged among the enlisted ranks in the 1990s because of poor pay, poor billets, the domination of barrack blocks by gangs, the absence of a professional noncommissioned officer corps, and *dedovshchina*, a hazing tradition that made a new Russian recruit 80 percent likely to be beaten up and 5 percent likely to be raped. See Dick, "A Bear Without Claws," p. 5.

[55]The average Russian infantryman received very little MOUT training. Russian MOUT tactical training for small units consisted of about 5 hours out of 151 total required. See Kostyuchenko, "Grozny's Lessons."

[56]According to Major Gregory Celestan, the present Russian system creates sergeants by taking raw recruits and training them for several months. See Gregory Celestan, *Wounded Bear: The Ongoing Russian Military Operation in Chechnya*, Fort Leavenworth, KS: U.S. Army, Foreign Military Studies Office, August 1996.

[57]For example, during the preparation for the assault on Grozny, no mockups of the city or its individual blocks were used.

[58]The Russian army was short 12,000 platoon leaders in 1994. Urban warfare and counterinsurgency operations place a heavy premium on small-unit commanders. See "The Chechen Conflict: No End of a Lesson?" *Jane's Intelligence Review—Special Report*, September 1, 1996; see also Zakharchul, "View of a Problem."

the Chechen defenders outnumbered the attacking Russians 15,000 to 6,000.

Most importantly, though, it was the Russian government's lack of a political-military strategy that integrated the seemingly disparate elements of the media, PSYOP, and ROE that cost them the war. The success of the Chechen political-military strategy probably serves as a wake-up call for future U.S. adversaries around the world.

## Table 3

### Battle Statistics for Recent MOUT

| Conflict | Strength of U.S. or Russian Forces at Start of Conflict | Strength of Enemy | U.S. or Russian Dead | U.S. or Russian Wounded | Enemy Dead | Enemy Wounded | Noncombatants Killed | Force Ratio |
|---|---|---|---|---|---|---|---|---|
| Firefight in Mogadishu | 140[a] + relief convoys | ≥2,000 | 18 | 73 | 500 | 814–1,000 | <300 | 1:14 |
| OJC | 26,000 | 4,000–6,000[b] | 23 | 324 | 314 | 124 | 200–300[c] | 5:1 |
| Battle for Grozny I | 6,000–8,000[d] | 10,000–15,000[e] | 1,100–8000[f] | 5,000–6,000[g] | 3,000–6,690[h] | ? | 24,000–25,000[i] | 1:2 |
| Battle for Grozny II | 12,000 | 3,000 | 500–1,000 | 1,407 (182 MIA) | ? | ? | ? | 4:1 |
| Chechen War | 20,000–40,000[j] | 15,000[k] | 6,000 | >13,000[l] | 2,000–3,000 | ? | 25,000–80,000[m] | 2:1 |

NOTE: This table presents approximate figures only. Both sides tended to distort casualties figures for propaganda purposes, so it is extremely difficult to make precise estimates. The reader should also note that strength numbers are given for the start of the conflict. Actual strength varied over the course of each conflict, especially in the Battle for Grozny I, when considerable Russian reinforcements were later sent into battle.

[a] This approximate figure includes the helicopter crews with the assault force of about 75 Rangers and 40 Delta Force troops.

[b] The PDF contained about 15,000 men, with an effective combat strength between 4,000 and 6,000. McConnell (1991), p. 30, and Cole (1995).

[c] Donnelly, Roth, and Baker (1991), p. 390, and Taw (1996), p. 8.

[d] One writer estimates that federal forces numbered 8,000 by February 1, 1995. See Korbut (1999).

[e] A recent estimate places Russian troop strength at 10,000. See Mukhin and Yavorskiy (2000).

fThis is an estimate of casualties up to February 10, from Novichkov et al. (1995). Mukhin and Yavorskiy (2000) estimate 1,500 were killed.

gSee Novichkov et al. (1995) and Grishin (1996).

hSee Novichkov et al. (1995).

iTwo sources conclude that 24,000 civilians were killed in Grozny through March 1995: see "Russia Pounds Rebel Positions Outside Capital of Chechnya," *The New York Times*, May 21, 1995, and "Russia's War in Chechnya: Urban Warfare Lessons Learned 1994–96," *Marine Corps Intelligence Activity Note*, CBRS Support Directorate (MCIA-1575-xxx-99), November 1998, p. 3. Others estimate 25,000; see Novichkov et al. (1995).

jInitially Grachev used a force of about 23,800 men, 80 tanks, 208 APC/IFVs, and 182 artillery pieces. Later, the Russian force grew to 38,000 men, 230 tanks, 454 APC/IFVs, and 388 artillery pieces. Lieven (1998, p. 280) believes about 40,000 Russian troops entered Chechnya, but because many Russian units were seriously understrength, the number may have been as low as 20,000.

kSee Raevsky (1995), who also cites Russian sources indicating that 10,000 Chechens were waiting in Grozny. Thomas (1995, p. 30) cites Russian estimates of 11,000–12,000 Chechens.

lThis is an estimate of Russian MOD wounded as of August 30, 1996. See Grishin (1996).

mThe Russian national security advisor, Alexander Lebed, estimated that 80,000 civilians were killed in the fighting in Chechnya. In contrast, journalist Anatol Lieven (1998, p. 108) believes the Lebed figures are exaggerated and that perhaps only 25,000 Chechen civilians died in the entire war.

# Chapter Three

# CROSS-CASE ANALYSIS

The premise of this monograph is that recent significant change in urban operations has more to do with information-related factors than with "traditional" military force factors.[1] This is not to say that information-related factors such as PSYOP or public affairs are now as decisive as a "traditional" MOUT factor such as airpower or combined arms teams. Killing enemy troops will probably remain the most efficacious way to defeat the enemy's will to fight. However, the marginal return from leveraging an information factor—such as the media—may be greater than the marginal return of applying more firepower.

To make a comparison between old and new, we need a baseline set of cases from which to start. This chapter begins by outlining the lessons learned from 22 battles fought before 1982, as described in *Modern Experience in City Combat.*[2]

---

[1]For example, airpower is an important factor in MOUT and it has changed, but not significantly in relative terms. Helicopters and PGMs are new to MOUT, but they have not been decisive. The media, on the other hand, has significantly changed enough that its role in recent MOUT has been qualitatively different than in the past. "Significant" change here means that the change in the MOUT factor is decisive enough to merit closer attention.

[2]See R. D. McLaurin, Paul A. Jureidini, David S. McDonald, and Kurt J. Sellers, *Modern Experience in City Combat*, Aberdeen Proving Ground, MD: U.S. Army Human Engineering Laboratory, March 1987.

## SUMMARY OF LESSONS BASED ON EARLIER MOUT

The influential publication *Modern Experience in City Combat* offers a baseline set of MOUT factors to start from.[3] Its analysis identified "the dominant factors historically affecting the course" of 22 selected urban battles that occurred between 1942 and 1982.  Table 4 lists the battles.  As the authors note, the careful selection of these 22 battles made their dataset too small to make unassailable conclusions, but it was big enough to vary some important parameters.  For example, they wanted their cases to cover a variation of attacker and defender victories, large and small cities, limited and general wars, duration of conflict, and the presence of air and naval support.  Thirty-two percent of the cases occurred during World War II; 45 percent took place between 1975 and 1982.  The authors looked for cases that included the employment of airpower by at least one side, large cities, and at least battalion-strength engagements.

Some of the main points of the report were the following:

- American forces should avoid cities where it is feasible.

- An attacker should encircle and isolate a city when possible.

- Airpower's important role is to cut off the city defenders from sources of supply and reinforcements.

- Armor has a definite role in MOUT.  Armor and APCs must have dismounted protection, however.

- Self-propelled artillery can be used to great effect as a direct-fire weapon in close combat.

- Airpower and artillery can have a positive psychological effect.

- The defender has a "good chance to win or at least prolong the battle and raise the cost for the attacker" if casualties and/or collateral damage can be limited.

- Combined arms operations have the best chance of success, especially when armor, infantry, and artillery train and develop doctrine together.

---

[3]*Modern Experience in City Combat* was intended to update lessons learned about MOUT from as recent a period as possible at the time of writing (1987).

## Table 4

### Baseline MOUT Cases

| Battle | Year | Force Ratio (attacker: defender) | Duration of Combat (days) | Limited or Unlimited? | "Winner" |
|---|---|---|---|---|---|
| Stalingrad | 1942[a] | 2:1 | >30 | U | Defender |
| Ortona | 1943 | 3:1 | 6–13 | U | Attacker |
| Aachen | 1944 | 1:3 | 14–30 | U | Attacker |
| Arnhem | 1944 | 1.5:1 | 6–13 | U | Defender |
| Cherbourg | 1944 | 3:1 | 6–13 | U | Attacker |
| Berlin | 1945 | 4.5:1 | 14–30 | U | Attacker |
| Manila | 1945 | 2.5:1 | 14–30 | U | Attacker |
| Seoul | 1950 | 3:1 | 6–13 | U | Attacker |
| Jerusalem | 1967 | 1.5:1 | 2–5 | L | Attacker |
| Hue | 1968 | 4:5 | 14–30 | U | Attacker |
| Quang Tri I | 1972 | 3:1 | 6–13 | U | Attacker |
| Quang Tri II | 1972 | 3:5 | >30 | U | Attacker |
| Suez City | 1973 | 1:5 | <1 | U | Defender |
| Ban Me Thout | 1975 | 7.5:1 | 1–2 | U | Attacker |
| Beirut I | 1975 | 5:3 | >30 | L | Draw |
| Tel Zaatar | 1976 | 1:1 | >30 | L | Attacker |
| Ashrafiyeh | 1978 | 10:1 | >30 | L | Defender |
| Khorramshahr | 1980 | 4:1 | 14–30 | U | Attacker |
| Zahle | 1981 | 15:1 | >30 | L | Defender |
| Beirut II | 1982 | 3:1 | >30 | U | Attacker |
| Sidon | 1982 | 4:1 | 2–5 | U | Attacker |
| Tyre | 1982 | 4:1 | 2–5 | U | Attacker |

[a]August to November only.
SOURCE: McLaurin et al. (1987), p. 94.

- Planning and intelligence are crucial to the outcome. Most defender "wins" were due to attacker intelligence failures.

- Preparation of the city was probably most critical for defender success.

- In no single case did casualties in the city itself alter the campaign outcome.

Overall, the 22 cases did not suggest any clearly emerging patterns in MOUT. Table 5 summarizes the major factors from the *Modern Ex-*

*perience in City Combat* research and compares them to this monograph's conclusions.

## THE CHANGING FACE OF URBAN OPERATIONS

The political environment of urban operations has changed in several ways in recent years.  Just as nuclear weapons introduced new limitations on the use of force after World War II, recent changes in the media, political justification, a growing abhorrence of violence, and evolving standards of morality have increased the restraints on the use of military force in urban operations today.[4] For the United States, military operations are now characterized by greater concern over public opinion, casualties of all sorts (including friendly, noncombatant, and even enemy casualties), and humanitarian issues.

News reporters are present on the battlefield in greater numbers than ever before.[5]  Peace operations in cities are particularly easy for reporters to gain access to.  In addition, because of the proliferation of smaller, more portable media devices, information technology is altering the political landscape of the battlefield.[6] Violence must be applied in a more discriminate manner because even the most minor

---

[4]The abhorrence is at least felt by the people of advanced market democracies.  In the modern postindustrial age, life expectancies are up, even the middle class is enjoying unprecedented prosperity, and war is increasingly considered barbaric and uncivilized.  Young men are avoiding the military and opting for the less rigorous life of an increasingly productive economy.  Recent "Nintendo" wars such as the Persian Gulf War have led to unrealistic expectations that war no longer has to be bloody.  Some scholars observe that the norms governing attacks on cities have evolved substantially since World War II, especially with the additional restrictions contained in the Additional Protocols of 1977 to the Geneva Conventions of 1949.  See Matthew C. Waxman, "Siegecraft and Surrender:  The Law and Strategy of Cities as Targets," *Virginia Journal of International Law,* Virginia Journal of International Law Association, Vol. 39, Number 2, Winter 1999, pp. 400–406.

[5]Charles Rick notes that only nine civilian war correspondents were present on the Island of Tarawa in the South Pacific in 1943 and fewer than 30 on the beaches of Normandy in 1944.  "The 600 reporters in the entire Pacific Theater in World War II were nearly matched by the 500 journalists who quickly appeared on tiny Grenada and in Panama City."  See Charles Rick, *The Military–News Media Relationship: Thinking Forward,* Carlisle Barracks, PA:  U.S. Army War College, 1993, p. vi.

[6]Information technology includes data processing and telecommunication technologies.

Table 5

MOUT Factors from *Modern Experience in City Combat* Viewed from Today's Perspective

| Factor | Past Description and Conclusions (based on 22 cases) | Present Conclusion (based on 3 cases) |
|---|---|---|
| *Intelligence* | A major consideration, usually attacker lost because of intelligence failures. | No significant change, still a crucial but elusive factor; HUMINT still more important. |
| *City size and composition* | Dictates location of defensive strongpoints. | Cities generally larger, operations more likely to be conducted in Third World shantytowns. |
| *Airpower* | Important for interdicting supplies and reinforcements into urban area; bombing ineffective. | No significant change—best used in more open areas or in isolating smaller towns and villages; weather still a factor; PGMs still not decisive; proliferation of RPGs and man-portable SAMs now a threat to rotary aircraft. |
| *Force size, force ratio* | Insufficient troop strength prevents encirclement of the city; if 4:1 advantage exists, attacker can win within two weeks on average. | Superior force ratio still crucial to encirclement but it is more elusive given the size of modern cities and conventional armies; generally less than 4:1 (attacker/defender) in present cases. |
| *Surprise* | Heavily linked to intelligence; tactical surprise by attacker can preempt defensive preparation. | No significant change—surprise does not seem to be more or less likely now. |
| *Offensive tactics* | Combined arms teams (infantry mixed with armor, artillery, or engineers) used with great success. Isolation of the defender rarely achieved completely, but it usually led to success. | No significant change—combined arms teams still the most effective force in MOUT. Isolation of the defender still important but unlikely in high-intensity MOUT because of the growth in city size and ongoing force structure cuts; not relevant to surgical and precision MOUT. Tactics have changed with regard to the use of noncombatants. |

**Table 5—continued**

| Factor | Past Description and Conclusions | Present Conclusion (based on 3 cases) |
|---|---|---|
| Defensive tactics | Mobile defense most successful; however, most MOUT centered around defensive strongpoints on avenues of approach. Defender reentry into cleared buildings was effective. Preparation of the city is most important. | No significant change—"swarming" anti-tank/anti-aircraft teams a new threat; tactics have changed with regard to the use of noncombatants. |
| Armor and artillery | Need infantry protection; best for isolating cities and direct-fire role. Both armor and artillery decisive in earlier battles (1945–1967) and less so in later period (1968–1982). | No significant change—more lethal man-portable anti-tank weapons and swarm tactics now a threat. ROE now prohibit armor and artillery in some cases. |
| ROE or "constraints" | Present in some battles, especially for cultural reasons (Jerusalem); defender has at least an equal chance to win if limitations to friendly or noncombatant casualties exist. | Significant change—ROE restrict the use of combined arms teams and airpower; ROE generally more restrictive because of the presence of the media and noncombatants |
| Noncombatants | In most cases civilians managed to evacuate battle zones or they were ignored. Civilians were used as hostages on occasion (Manila, Sidon, Tyre, Beirut II). In no single case did casualties in the city itself alter the overall campaign outcome. | Significant change—noncombatants now used for human shield tactics, intelligence, cover, and concealment. |
| Media | Not critical; noted, however, for contribution toward the "strategic implications" of urban operations. | Significant change—media now a PSYOP tool, part of an integrated political-military strategy to erode U.S. popular support. |
| Public affairs | Not identified as an important factor. | Significant change—influenced the outcome in all three cases. |

**Table 5—continued**

| Factor | Past Description and Conclusions | Present Conclusion (based on 3 cases) |
|---|---|---|
| Civil affairs | Not identified as an important factor. | Significant change—influenced the outcome in all three cases. |
| Political-military strategy[a] | Whether a battle was a win or not depends on the objective one is concerned with. Political constraints—such as ROE—can restrict tactics. | Significant change stemming from the synergies of information-related factors such as PSYOP, civil affairs, public affairs, ROE, and control of the media and noncombatants. |
| Information operations (PSYOP, IW, EW) | Not identified as an important factor. | Significant change—influenced the outcome in all three cases. |

NOTES:  McLaurin et al.'s factors are in italics.  Five additional factors are added here for consideration:  media, public affairs, civil affairs, political-military strategy, and information operations.  The factors in the shaded rows have undergone the most change in the last decade (based on the three cases in this monograph).

A few other factors from *Modern Experience in City Combat* are not listed here—such as force structure, weather, and proximity to trafficable waterways—because they were not elaborated upon in that report or are subsumed by other factors already listed.

[a]*Modern Experience in City Combat* lists the closest comparable factors as "role of the battle in the campaign" and "objectives."

acts of violence can be broadcast to millions of voters.[7] The more people with portable commercial equipment, the greater the chance that battlefield drama will be recorded. Political constraints on the use of military force have increased because democratically elected leaders are loath to expose voters to the brutal images of war. Today, uncensored information can be provided to the public in near-real-time, video form.[8] Video footage of the mutilated, naked American corpse being dragged through the dusty streets of Mogadishu in October 1993 serves as one example of a media event that prompted a public outcry.

There seems to be a greater concern over noncombatant casualties than in the past, especially when the media are present.[9] Tolerance levels are changing because the new weapons are believed to be more surgical. Adversaries have tried to capitalize on this sensitivity to bloodshed. The human shield tactics witnessed recently in Iraq and the Balkans prevented the use of airpower when civilians positioned themselves on strategic targets like bridges. When NATO bombs hit a convoy of refugees in Kosovo in 1999, some of the first Serbs on the scene were armed with cameras.

War is now sometimes justified on moral or humanitarian grounds rather than serious national security interests.[10] For example, in

---

[7]The cumbersome television satellite equipment which had to be transported on aircraft pallets to Panama in 1989 can now be carried in a few small cases (Rick, *The Military–News Media Relationship*, p. 15). The equipment needed for a live feed can now be handled by a two-man crew carrying less than 100 pounds in two cases (digital camera, wideband cellular phone, satellite dish, and laptop computer). See Captain Scott C. Stearns, "Unit-level Public Affairs Planning," *Military Review*, December 1998–February 1999, p. 24. Also, the proliferation of cheap digital movie-making technology is creating more opportunities for information warfare and deception. For a total of about $4,000, a combination of a new digital camcorder, special software, and a mid-range PC puts the power to make VHS-quality movies in the hands of the general population.

[8]One wonders whether the Vietnam War might have ended sooner if all recent telecommunication advances—digital camcorders, digital satellite phones, faxes, and commercial imaging satellites—had been present in the 1960s. How many of those 50,000 American casualties would have been tolerated before political pressure brought the war to an earlier halt? What if millions of Americans had been able to download and play a video of the My Lai massacre on their home computers?

[9]Video and still images seem to increase the shock value of violence.

[10]The most recent grand strategy statement by the White House in December 1999, *A National Security Strategy for a New Century,* lists three types of national interests:

March 1999, President Clinton announced that Operation Allied Force, the NATO attack on Yugoslavia, was launched because the United States had a "moral imperative" to save the ethnic Albanians from Milosevic's ethnic cleansing campaign. This more altruistic concept of national interest has been called "the Clinton Doctrine."[11] There is a growing body of international law that permits armed intervention for humanitarian purposes even without specific UN approval.[12]

When military action is conducted for less-than-vital national security interests, political support at home may be more fragile and susceptible not only to casualties but also to enemy information operations.[13] Humanitarian missions are generally prolonged interventions in complex political environments characterized by civil conflict, where U.S. interests are less compelling, if they are clear at all. Studies have shown that the U.S. public is willing to accept loss of life only if the interests and values are judged important enough.[14] Operations built upon tenuous political-military links—

---

vital interests (vital to national survival), important national interests (which affect the character of the world in which we live), and humanitarian and other interests. Official policy clearly states that military force is justified if "our values demand it." See page 6 of the document.

[11]It remains to be seen whether future administrations will be willing to commit U.S. military force for humanitarian purposes.

[12]International law consists of provisions of the UN Charter, treaties, and activities and practices that have won broad acceptance over the years. Norman Kempster, "Leaders and Scholars Clash Over Legality," *Los Angeles Times,* March 26, 1999.

[13]One illustrative example is Operation Allied Force (OAF) in 1999. Recognizing that political support is more sensitive to casualties when military action is conducted for less-than-vital national security interests such as a "moral imperative," the Serbs sought to raise the human and moral costs of conflict in order to erode the will of the American people. They tried to raise the human cost by inflicting as many American casualties as possible; at the same time, by increasing the number of noncombatant deaths from NATO bombs, they tried to undermine NATO's moral justification for the use of force.

[14]For example, Larson reports that support for the humanitarian operation in Somalia fell 10 points after the firefight in October 1993 (it had already declined 35 points even before the fight). In contrast, public support for the invasion of Panama remained high even after casualties were incurred because of President Bush's argument that Americans were in danger in Panama. See Larson, *Casualties and Consensus,* pp. 41, 50, 71.

low-value political goals that require high costs—are vulnerable to enemy strategies aimed at domestic public opinion.[15]

The recent MOUT cases in this study may also reflect a larger trend in the nature of war—that is, armed conflict is more likely to involve low-intensity forces because the spread of weapons of mass destruction deters high-intensity conventional war.[16] If this is true, the wars of the future will probably look more like the Mogadishu firefight and less like the desert tank battles of the Persian Gulf War. Third World conflicts usually involve additional political constraints on the use of military force.[17] The risk of lengthy stalemate is higher in low-intensity conflicts, so mounting casualties tend to serve as a lightning rod for public dissatisfaction.[18]

---

[15]In the Somali case, the benefits were never perceived by most to have warranted much loss of life. 60 percent of those polled by *Time*/CNN on October 7, 1993, agreed with the statement that "Nothing the United States could accomplish in Somalia is worth the death of even one more soldier." See Larson, *Casualties and Consensus*, p. 47.

[16]One of the scholars who argue this is Martin Van Creveld. In his book, *The Transformation of War*, he argues that the use of armed force as an instrument for attaining political ends by major states is less and less viable because of the presence of nuclear weapons. Although the book was published at an unfortunate date (just before the onset of the Persian Gulf War), it does raise several telling points. In every volatile region where conventional wars used to be fought (such as the Middle East, South Asia, and China's periphery), the introduction of nuclear weapons has coincided with a marked decline of conventional war. The new dominant form of war is low-intensity conflicts (LICs). Since 1945, about three-quarters of the 160 armed conflicts worldwide have been nonconventional or of the "low-intensity" variety. Van Creveld argues that LICs have also been more politically significant than conventional wars, in terms of both casualties and territorial boundaries. What's more, the major states involved have lost the vast majority of these wars. Because conventional military power—high-tech tanks, artillery, airpower, etc.—is all but useless against insurgents, he hypothesizes that the rise of LIC will render the military forces of major states irrelevant.

[17]Constraints have shaped and limited U.S. policy and strategy in the Third World since the start of the nuclear era. One analysis of the Korean War, the Cuban Missile Crisis, and the Vietnam War concluded that U.S. constraints were motivated by several concerns: to avoid direct military conflict with the USSR, to avoid friendly and enemy civilian casualties, to limit U.S. casualties, and to accommodate U.S. allies. See Steve Hosmer, *Constraints on U.S. Strategy in Third World Conflict*, Santa Monica, CA: RAND, R-3208-AF, 1985.

[18]On the basis of poll data and extensive interviews, Mark Lorell and Charles Kelley concluded that casualties were the single most important factor eroding public support in limited wars in the Third World. See Mark Lorell and Charles Kelley, *Casualties, Public Opinion, and Presidential Policy During the Vietnam War*, Santa Monica, CA: RAND, R-3060-AF, 1985, p. vii.

Insurgent forces generally seek to avoid warfare on open ground where the airpower and other sophisticated weapon systems of the United States can be brought to bear. Urban operations are one way to do this. The urban environment offers not just physical cover and concealment but also political cover behind noncombatants. By seeking to inflict as many casualties as possible, the weaker state can follow an asymmetric strategy that concentrates on subduing the will to fight of the American people rather than defeating American military forces.[19] The classic guerrilla strategy—to win by not losing— can create the impression that U.S. forces are fighting in a quagmire, which diminishes the prospects for success in the eyes of the public.

In short, all of these political, technological, and social developments increase the importance of information operations (and related activities) during urban operations (see Figure 3). Information operations focus on the perception and will of the people fighting the war, the support of the domestic population at home, as well as the support of the indigenous population in the urban operations theater. More opportunities exist than ever before to subdue the will of the enemy through information manipulation (in addition to destroying his military forces).

The geostrategic problem for the United States is to figure out how to (1) subdue the will of the enemy in conflicts involving less-than-vital interests while (2) maintaining popular support from the American people. The former can be achieved by killing the enemy and by controlling information. The latter can be achieved by minimizing casualties, exercising political leadership, and controlling information.

Before proceeding further, the official doctrinal language of information operations (IO) should be outlined and defined.[20]  For the

---

[19]During OAF, even the common Serb on the street realized that the objective was to raise the cost of military action beyond the U.S. public's threshold of tolerance. As one Serb said, "Clinton didn't succeed in Somalia when they were killing Americans on the street. We will do the same. The people who fall from the plane: We will find them." See David Holley, "Serbs Rally Around Their Leader," *Los Angeles Times*, March 26, 1999.

[20]These are Joint Staff and Army definitions. See *Joint Doctrine for Information Operations*, Joint Pub 3-13, 9 October 1998; *Joint Doctrine for Command and Control Warfare (C2W)*, Joint Pub 3-13.1, 7 February 1996; *Doctrine for Joint Psychological*

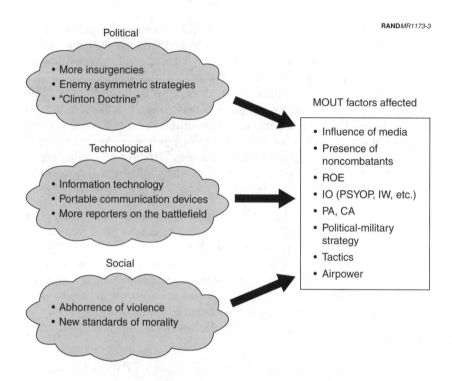

Figure 3—The Changing Environment of Urban Operations

purposes of this discussion, information operations involve actions taken to affect the adversary's information and information systems and to defend one's own.[21]  Ultimately, IO is designed to influence the enemy's emotions, motives, reasoning, and behavior.  IO at the strategic level of war includes influencing all elements of an adversary's national power (military, political, economic, and informa-

---

*Operations,* Joint Pub 3-53, 10 July 1996; and *Public Affairs Operations,* Field Manual (FM) 46-1, Department of the Army, 30 May 1997.

[21]A subset of IO is information warfare (IW).  IW is information operations during a time of crisis designed to achieve specific goals over a specific adversary.  A subset of IW is command and control warfare (C2W).  C2W is an application of IW in military operations that specifically attacks and defends command and control targets.

tional).[22]  At the operational level, IO focuses on lines of communication, logistics, and command and control to achieve campaign objectives.  Tactical-level objectives are met through IO attacks on adversarial information-based processes directly related to the conduct of military operations.[23]

The basic components of offensive IO include psychological operations (PSYOP), electronic warfare (EW), physical attacks, deception, special information operations (SIO), and operational security (OPSEC) (see Figure 4).[24]  Public affairs (PA) and civil affairs (CA) are *information-related* activities.[25]

PSYOP are actions taken to convey selected information to foreign audiences.  PSYOP targets the will and morale of enemy combatants and noncombatants and may support military deception.  A classical example is to drop propaganda leaflets over target populations.  EW is any military action involving the use of electromagnetic and directed energy to control the electromagnetic spectrum or attack the enemy.  Physical attack is self-explanatory.  SIO are information operations that, by their sensitive nature, require a special review and approval process.  OPSEC denies the adversary critical information

---

[22]Some authors have postulated that another way information exerts power today is at a strategic-cultural level.  Joe Nye calls this "soft power," the power that cultural influences have on foreign populations.  "Soft power" is co-optive power, or the ability of a country to structure a situation so that other countries develop preferences or define their interests in ways consistent with their own.  Political leaders have long understood the power of attractive ideas or the ability to set the political agenda and determine the framework of debate in a way that shapes others' preferences.  The rest of the world indirectly conforms to American ideals because of the globalization of American culture (American films, for example, account for only 6–7 percent of all films made but occupy about 50 percent of world screen time) and the U.S. monopoly on many aspects of the information revolution (in 1981 the United States was responsible for 80 percent of worldwide transmission and processing of data).  See Joseph S. Nye, Jr., and William A. Owens, "America's Information Edge," *Foreign Affairs,* Vol. 75, No. 2, March/April 1996, p. 21.

[23]For our purposes, the discussion will concentrate on the more strategic applications of IO that influence populations and national will, not necessarily C2W actions concerned with disrupting C2 systems.

[24]Defensive IO primarily protect and defend information and information systems.  Defensive IO activities include information assurance, OPSEC, physical security, counterdeception, counterpropaganda, counterintelligence, EW, and SIO.  See *Joint Doctrine for Information Operations,* Joint Pub 3-13, October 9, 1998, for more details.

[25]Normally a Joint Force Commander would set up an IO cell that contains representatives from all the above elements.

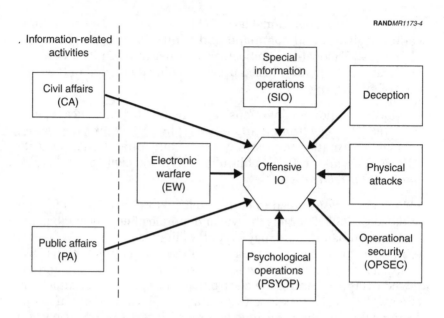

**Figure 4—Information Operation Components and Related Activities**

about friendly capabilities and intentions needed for effective decisionmaking. Military deception targets adversary decisionmakers
through effects on their intelligence collection, analysis, and dissem-
ination systems. Deception induces misperception; ultimately, the
target is "the human decisionmaking process."[26]

Public affairs and civil affairs are related activities that target the U.S.
population (and media) and indigenous population respectively. PA
keeps the U.S. public and armed service personnel informed as to
military goals and current operations while countering any disinformation spread by the enemy. The PA motto is maximum disclosure
with minimum delay. CA encompasses activities that a commander
takes to establish relations with civil authorities and the general
population where his forces are deployed. CA and PA both complement PSYOP.

---

[26]See Scott Gerwehr and Russell Glenn, *The Art of Darkness: Deception and Urban
Operations,* Santa Monica, CA: RAND, MR-1132-A, 1999.

All of these IO-related elements may be more effective in future urban operations because of the political, social, and technological developments described earlier. The "traditional" factors drawn from urban operations in the past—intelligence, armor, airpower, etc.—remain crucial for the goal of killing the enemy and minimizing U.S. casualties. But the factors crucial to information operations—ROE, PSYOP, public and civil affairs, information warfare, and a political-military strategy that integrates these efforts—are growing in significance and deserve more attention. This is especially true for counterinsurgency operations that aim to gain the support of the local population. For example, it may be possible to persuade a city population to stop supporting indigenous soldiers (and even expel them, as the citizens of Gudermes in Chechnya did in November 1999).

Influence charts might help the reader visualize these seemingly disparate elements. Figure 5 is a simple influence chart that shows the framework through which both physical attacks and information attacks can affect the will to fight (shown from a U.S. perspective). It is one snapshot to illustrate how information manipulation might occur. One can picture the process as a flow.

A political-military strategy must consist of *goals*, a *means* to achieve them, and *ways*, a plan or a method for applying the means. Goals that are explicitly defined and justified for the public help stabilize domestic support in the face of casualties. Polling data show that the public becomes less tolerant of casualties when the prospects for success are low, when the perceived benefits do not justify high costs, or when there is a lack of consensus among political leaders.[27] Political consensus over policy leads to more favorable media coverage. Indeed, media reporting is often *indexed* to the tone of the leadership debate—in other words, media reporting will generally be favorable if most leaders and experts support a policy, and negative if they are critical of the policy.[28]

---

[27]See Larson, *Casualties and Consensus*, pp. xv–xviii.

[28]See Daniel L. Byman, Matthew C. Waxman, and Eric Larson, *Air Power as a Coercive Instrument*, MR-1061-AF, 1999, p. 69.

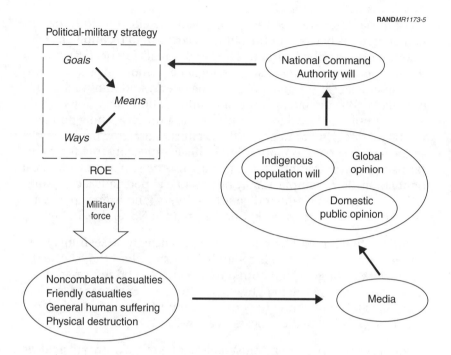

**Figure 5—Influence Chart of Political-Military Links**

When domestic political constraints are incorporated into the political-military strategy, ROE result. ROE shape how military means are applied, which in turn influences the number of friendly and noncombatant casualties and collateral damage. More restrictive ROE can increase the risk of friendly casualties.[29] In every mission, it is important to ask what the ROE are and whether the

---

[29]For example, ROE during Operation Deliberate Force increased the risk to pilots. Special instructions were issued to aircrews, for example: (1) those attacking a bridge must make a dry pass over the target and attack on an axis perpendicular to it, releasing only one bomb per pass. (2) Those carrying out suppression of enemy air defense (SEAD) strikes were not authorized without special approval to conduct preemptive or reactive strikes against surface-to-air missile sites except under certain restrictive conditions. See unpublished manuscript by Alan Vick, John Stillion, David Frelinger, Joel S. Kvitky, Benjamin S. Lambeth, Jeff Marquis, and Matthew C. Waxman, "Exploring New Concepts for Aerospace Operations in Urban Environments," November 1999, p. 60.

mission can still be accomplished with acceptable losses. In extreme cases, the ability of U.S. forces to overcome an opponent may be limited more by the political constraints embodied in ROE than by the enemy's military capability.[30]

The application of military force can result in noncombatant and friendly casualties, human suffering, and physical destruction, all of which are subject to media scrutiny. People are informed of these costs of war, the impact depending in part on the level of media access.[31] When events are closely monitored by the media, even minor tactical events can have strategic outcomes. There are compelling data showing that public support for war declines as friendly casualties increase.[32]

Media coverage of these costs of war and any attendant political debate influences U.S. public opinion, the will of the indigenous population in the theater of operations, and global opinion. A shift in public support can influence to some degree the national command authority's willingness to continue risking the lives of U.S. soldiers.[33] If the human costs of achieving the current military goals

---

[30]Brigadier General John R. Groves, "Operations in Urban Environments," *Military Review*, July–August 1998, p. 35.

[31]The indigenous population is directly affected by the use of military force of course.

[32]There is an extensive literature on this subject. John E. Mueller's *War, Presidents and Public Opinion* (1973) was one of the original studies that observed the log of cumulative casualties as the best predictor of public support (based on data from Korea and Vietnam). Gartner and Segura recently found marginal casualties to be the best predictor when casualties are increasing and the log of cumulative casualties the best predictor when casualties are decreasing. See Scott Sigmund Gartner and Gary M. Segura, "War, Casualties, and Public Opinion," *Journal of Conflict Resolution*, Vol. 42, No. 3, June 1998, pp. 278–300. A related study argues that casualties influence the duration and outcome of wars—see Scott D. Bennett and Allan C. Stam III, "The Declining Advantages of Democracy: A Combined Model of War Outcomes and Duration," *Journal of Conflict Resolution*, Vol. 42, No. 3, June 1998, pp. 344–366.

[33]For the purposes of this monograph, it is assumed that adverse effects on public support are at least weighed in the decisionmaking process as additional costs, just as the other costs of military action are weighed (such as friendly casualties, international opinion, collateral damage, etc.). The proposition that public support and opinion affect the decisionmaking of the national command authority is debated endlessly in the literature (for example, see Holsti). It seems logical to assume that in many cases—especially in short crises—foreign policy decisions are made independent of public opinion because of the requirements for secrecy, speed, and flexibility. Some studies conclude that public opinion is irrelevant because analysis of polling data from past conflicts indicates the public was poorly informed and their opinions were

outweigh the perceived benefits, domestic political pressure can possibly force a change in policy, an adjustment of ROE, or termination of an operation.[34] It is imperative that political-military strategy keep the human costs of combat—or the awareness of those costs—under a threshold of public tolerance.

This basic framework has not changed fundamentally, but the opportunities for IO and the ability to influence an opponent's will to fight are increasing. News also appears to travel much faster in the information age.

The influence of the media is potentially more powerful now because television coverage of wars is more extensive and noncombatants are more prevalent in urban environments.[35] The Persian Gulf War has been called the "mother of all media events": television transmitted 4,383 stories of the crisis over a seven-month period.[36] In the ever brighter media glare, an increasing presence of noncombatants on the battlefield is significant because the death of women and children can strike deep emotional chords with the public.

---

volatile and lacked structure and coherence. Since many institutions shape, mobilize, and transmit public sentiment, such as the media, special interest groups, and legislators, appropriate indicators of public opinion are sometimes not even readily apparent.

[34]For example, during the Persian Gulf War, pictures of 300 civilian dead in the aftermath of the U.S. bombing of an Iraqi bunker in Baghdad (which was also being used as an air raid shelter) led to future restrictions on bombing of targets. Jeremy Shapiro, "Information and War: Is It a Revolution?" in Zalmay M. Khalilzad and John P. White (eds.), *Strategic Appraisal: The Changing Role of Information in Warfare*, Santa Monica, CA: RAND, MR-1016-AF, 1999, p. 125.

[35]In World War II, the media consisted of print reporters like Ernie Pyle. Public access to the horror of war was limited, censored, delayed, and in the form of print and still images. The dirty underbelly of war—atrocities, mutilations, graphic carnage—was generally less visible. In the Vietnam era, there were no all-news cable channels. Live pictures of combat were unheard of because correspondents had to physically transport their film to the airport so it could be flown to New York. The newscast would appear two or three days later. In the 1990s, information was provided to the public in real time, in video form, and often uncensored.

[36]John E. Mueller, *Policy and Opinion in the Gulf War*, Chicago: University of Chicago Press, 1994, p. xiv. For comparison, just before and during the Tet offensive there were 187 television stories on the Vietnam War between September 1967 and January 1968, and 457 television network weekday evening news reports between January 29 and March 28, 1968. Only 118 of these were supplied by newsmen actually in Vietnam. See Peter Braestrup, *Big Story: How the American Press and Television Reported and Interpreted the Crisis of Tet 1968 in Vietnam and Washington*, Boulder, CO: Westview Press, 1977, p. 41.

"Media manipulation" is included in Figure 6, even though this remains a troublesome concept because it implies denial of the free press. Current doctrine states that PA officers should not manipulate public opinion but seek to disclose as much as possible as soon as possible.[37] Military commanders may have some control over media access, but this will be difficult in cities, and the more so during humanitarian operations. However, there are subtle and indirect ways in which the media may be influenced that go beyond the straightforward mission of public affairs units, without undermining the credibility of the military for honesty.[38] Press pools are useful for restraining reporters on the battlefield. Reporters can also be "inadvertently" delayed, steered away from certain areas, assigned to certain units, etc. The military can try to shape the public's perception of events by selectively releasing information to the media that promotes its agenda, such as video footage of high-precision bombs in action.[39] Extolling the virtues of high technology downplays the human costs of combat.

An effective political-military strategy integrates all the information tools available (PSYOP, PA, CA, and IW) and the media to influence the battle of wills. There are mutually reinforcing relationships—even synergy—between many of these elements. For example, coordination between public affairs, civil affairs, and psychological operations results in a focused message for managing the perception of the indigenous population in the area of operations. PSYOP and civil affairs units help remove noncombatants before a battle commences (thereby lowering possible noncombatant casualties) and increase HUMINT.[40] PA and CA units interact with the media.

---

[37]Without violating operations security, of course.

[38]Honesty is important because truth builds credibility with the target audience. See Major Mark R. Newell, "Tactical-Level Public Affairs and Information Operations," *Military Review*, December 1998–February 1999, p. 23.

[39]For example, during the Gulf War, images of Patriot missiles knocking Iraqi Scuds out of the nighttime sky over Tel Aviv created a public perception of the wonders of American military technology, persuaded the Israelis to refrain from attacking Iraq, and allayed the fears of the Israeli population. Subsequent studies demonstrated that the Patriot may have failed to hit a single target during the course of the entire war.

[40]HUMINT is more available if friendly forces can gain the support of the civilian population. For support and stability operations in particular, it is critical that the support of the indigenous population be targeted through the proper use of ROE, the media, and PSYOP. Roger Trinquier and others have argued that control of the popu-

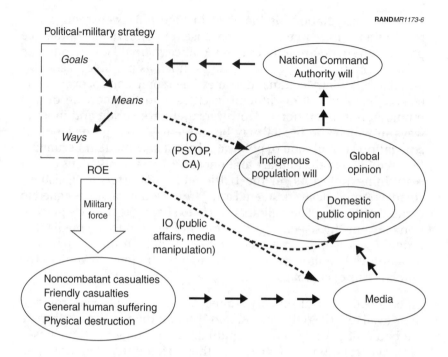

Figure 6—Influence Chart with IO

ROE can affect PA, CA, and PSYOP. Permissive ROE can precipitate civilian casualties, which attracts more media. Overly restrictive ROE cause friendly casualties. Some ROE—like graduated response approaches that use loudspeakers, warning shots, and firepower demonstrations—have PSYOP implications. IO tools can also maintain public support in the United States in the face of noncombatant casualties.[41]

---

lation can provide a significant advantage in urban warfare. Goligowski names several sources that recognize the importance of population control. See Goligowski, *Operational Art and Military Operations on Urbanized Terrain*, pp. 31–32.

[41]For example, in the Gulf War, the American public was mostly insensitive to Iraqi civilian casualties because they believed Saddam was to blame for placing military targets in civilian areas. The Bush administration effectively demonized Saddam and identified the important national security interests at stake. Seventy-one percent of those polled in February 1991 said the United States was justified in attacking military targets that Saddam had hidden in areas populated by noncombatants (*Los Angeles*

Because of the faster flow of events, a political-military strategy must also be adaptive, responding to the changing situation on the battle-field. In the city, commercial video of a firefight can reach television audiences before the military situation report (SITREP) works its way up the chain of command.

The Chechen War (1994–1996) provided a good example of how this political-military process works. A democratic state waged war for less-than-vital national interests and without the benefit of a political-military strategy focused on information operations. Per-missive Russian ROE and poor CA discredited pro-Moscow political movements inside Chechnya.[42] The Russians allowed the Chechen rebels to consolidate the support of the indigenous population. Russian PA was poor and management of the media was almost nonexistent.

The Chechens, for their part, used the media and noncombatants for PSYOP. They managed to lower the morale of the Russian army and undermine Russian domestic public support for the war—and they did this to a stoic people who historically have always been willing to make great sacrifices in war.

The Chechen army was inferior to the Russian military in terms of resources. Its best recourse was to defeat the will of the Russian people by raising the cost of winning the war to an unacceptable level.[43] The Chechens recognized the unique opportunities that an urban operations environment offered in that regard.

---

*Times,* February 15–17, 1991); 67 percent said they thought the United States was making enough effort to avoid bombing civilian areas (ABC News/ *Washington Post,* February 14, 1991). See Byman, Waxman, and Larson, *Air Power as a Coercive Instrument,* p. 78, and Mueller, *Policy and Opinion in the Gulf War,* p. xvii.

[42]At the start of the war only a fraction of the Chechen population was actively hostile to Russian forces. That fraction increased as death and destruction continued to rain down from above. As the Russian national security adviser Lebed said, "When we were entering that country, 90 percent of the population were welcoming us, lining the roads, flowers in their hands. When we were withdrawing from it, we were hated by everyone." Chechens who had lost a relative were especially bad: "They became wolves." See Gall and de Waal, *Chechnya: Calamity in the Caucasus,* p. 348.

[43]As Clausewitz observed, wars end when one side imposes its will on the other. That occurs when either the opposing army is physically destroyed or when the willpower of the population that supports the army is influenced to stop the war. Weaker opponents who cannot achieve the former must seek the latter result. See Carl von

In the cross-case analysis that follows, the premises noted above are validated by looking at several MOUT factors in detail.

## FACTORS UNDERGOING SIGNIFICANT CHANGE

Given the growing relevance of information operations, several factors appear to have grown in significance over the last decade: the presence of noncombatants, the presence of the media, ROE, PSYOP, IO-related activities such as civil affairs and public affairs, and political-military strategy.

### Presence of Noncombatants[44]

In recent urban operations, the presence of noncombatants significantly affected tactics, planning, ROE, and political-military strategy. Noncombatants were present in greater numbers, they played an active role in the fighting, they made ROE more restrictive, and they attracted the media.

There are a number of reasons why the number of noncombatants generally increased. Adversaries found cities a useful asymmetric avenue to face superior conventional armies. Insurgents utilized city dwellers for cover, concealment, and support. In the surgical and precision MOUT cases, there was usually no time or need for civilians to evacuate the combat zone. Even in the high-intensity case, many noncombatants remained despite the scale of destruction, and civilians wandered around Grozny throughout the fighting.[45]

An increase in the presence of noncombatants created the need for more restrictive ROE. Rules of engagement were needed because the

---

Clausewitz, *On War,* edited and translated by Michael Howard and Peter Paret, New York: Knopf, 1993.

[44]The standard definition of a noncombatant is a man, woman, or child who is not actively engaged in military-related activities and who is a civilian. Once a civilian actively engages in military activities he or she is considered a combatant according to the law of armed conflict. For the purposes of this monograph, civilian women and children are always referred to as noncombatants.

[45]About 300,000 Chechen civilians did flee Grozny during the fighting. Adam Geibel, "Lessons in Urban Combat: Grozny, New Year's Eve, 1994," *Infantry,* Vol. 85, No. 6, November–December 1995, p. 24.

indiscriminate killing of civilians provides a moral and psychological advantage to the enemy, erodes domestic and international support for the use of force, and strengthens the will to resist among the indigenous population. Also, in recent years, there has been a growing trend for victims of war to take legal action.[46]

Civilians impeded operations, especially when no discriminate or nonlethal means of force was available (or considered). During the initial stages of the Chechen conflict, Russian troops obeyed orders not to kill civilians. Because Russian soldiers lacked any nonlethal means of crowd control and their ROE were not clear, Chechen civilians were allowed to blockade resupply convoys and even set fire to Russian vehicles. Unarmed civilian crowds, mostly women, slowed or halted the advance of all three armored columns approaching Grozny in December 1994. Russian IFVs were taken and reportedly handed over to the Chechens.[47] Major General Ivan Babichev stopped his advance toward Grozny because he refused to "wrap bodies round the tracks of his tanks."[48]

In Panama, the presence of civilians in the residential areas of Quarry Heights and Albrook Air Station required new techniques for the application of force. To try to minimize casualties and collateral damage, U.S. troops used "graduated response." First they used loudspeakers to entice the defenders into giving up without a fight. Then they put on a demonstration of AC-130 firepower nearby, threatening to move that destructive firepower onto the Panamanian position if they did not surrender immediately. The PDF soldiers either surrendered or fled.

Noncombatants played a significant role in the actual fighting during recent urban operations, especially when the conditions were right (i.e., an insurgency environment in which the population is hostile

---

[46]The case of a Panamanian woman who was killed by the collateral damage of a 2.75-inch Cobra rocket became a symbol of a campaign for financial compensation for Panamanian civilian casualties. Holocaust victims have settled with the Swiss government. German companies are currently being sued in U.S. courts for their use of slave labor during World War II. The financial cost of noncombatant deaths could be substantial in the future.

[47]See Lieven, *Chechnya: Tombstone of Russian Power*, p. 103.

[48]See Raevsky, "Russian Military Performance in Chechnya: An Initial Evaluation," p. 684.

from the very start and ROE prevents the indiscriminate slaughter of civilians). Noncombatants were used for fighting, cover and concealment, and situational awareness. In these roles, noncombatants served as a useful tactical asymmetric response to superior U.S. conventional forces.

For example, during the October 1993 firefight in Mogadishu, Somali noncombatants participated directly in the fighting as fighters or scouts, or indirectly as a sort of mobile screen for Somali fighters. Armed Somalis deliberately used noncombatants for cover and concealment because they knew the Americans had been issued strict rules of engagement. Rangers were under orders to shoot only at people who pointed a weapon at them.[49]

Noncombatants posed a major problem for conventional forces because they enabled the enemy to move like—to use Mao's analogy— "fish swimming in the sea."[50] For example, Chechen snipers attacked Russian soldiers and then donned Red Cross armbands to mingle with civilians and conceal themselves.[51] In Chechnya, it was often impossible to distinguish between noncombatants and combatants because they wore similar attire. Somali gunmen found it easy to blend into gathering onlookers, using noncombatants as cover while they moved their forces toward the helicopter crash sites. The fact that none of the clans wore uniforms or other distinctive clothing helped conceal them among noncombatants.

The practice of firing from behind women and children and using them for mobile cover and concealment was standard operating procedure for the Somalis.[52] As a result, about a third of all Somali

---

[49]At one point, a Ranger saw a Somali with a gun prone on the dirt between two kneeling women. He had the barrel of his weapon between the women's legs, and there were four children actually sitting on him. He was completely shielded by noncombatants. See Bowden, *Blackhawk Down*.

[50]See David Miller, "Big City Blues," *International Defense Review*, Vol. 28, Issue 3, March 1, 1995.

[51]Russian soldiers at checkpoints countered this tactic by stripping the shirts off of suspected Chechen males and looking for telltale signs of a soldier, such as rifle recoil bruises on the shoulder, gunpowder on the clothes or fingers, etc. See Thomas, "Some Asymmetric Lessons of Urban Combat."

[52]Even as early as March 1993, in Kismayo two clans used women and children as active participants, with a mix of carefully coordinated infantry tactics.

casualties in the firefight were women and children.[53] The Chechens deliberately placed artillery near schools and apartment buildings to discourage Russian attacks (many of the remaining civilians were ethnic Russians).[54] Dudayev placed his air defense ZSU-23/4s in residential areas. Pavel Grachev claims that Chechens used noncombatants as "human shield" cover when attacking from hospitals, schools, and apartment blocks.[55] During the raid on Budyonnovsk, Basayev used his hundreds of hostages in the hospital siege as cover. Chechens made hostages stand at the windows of the hospital so they could fire from behind them.[56]

Even if the civilian population was not hostile, noncombatants still offered cover and concealment. The OJC case fits this description—in general, the Panamanian people were not overtly hostile and U.S. troops faced no large-scale uprisings or popular resistance.[57] This lack of support for Noriega made OJC much easier, but it did not prevent some PDF soldiers from using noncombatant areas as cover. For example, during the air assault on Tinajitas, 82nd Airborne troops loaded on Blackhawk helicopters took fire from PDF snipers firing from crowds of civilians. Apaches, Cobras, and OH-58s could not prepare the landing zones because of nearby civilian neighborhoods. ROE prevented return fire because civilians were in the area.[58]

It should be noted that *both sides* may have used noncombatants in Somalia. Somali eyewitnesses have charged that Somali women and

---

[53]See John H. Cushman, "Death Toll About 300 in October 3 U.S.-Somali Battle," *The New York Times*, October 16, 1993.

[54]"Russia's War in Chechnya: Urban Warfare Lessons Learned 1994–96."

[55]Cited from FBIS report in Thomas, "The Caucasus Conflict and Russian Security: The Russian Armed Forces Confront Chechnya, III. The Battle for Grozny, 1–26 January 1995,"p. 56.

[56]Gall and de Waal note that the elite Russian Alpha snipers worked as a team to fire at hostage legs to drop them before taking out the Chechen gunman. See Gall and de Waal, *Chechnya: Calamity in the Caucasus*, p. 270. The Chechens also used human shields during the Kizlyar-Pervomaiskoye raid. See John Arquilla and Theodore Karasik, "Chechnya: A Glimpse of Future Conflict?" *Studies in Conflict and Terrorism*, Vol. 22, No. 3, July–September 1999, p. 220.

[57]See Jennifer Taw, *Operation Just Cause: Lessons for Operations Other Than War*, Santa Monica, CA: RAND, MR-569-A, 1996, p. vii.

[58]See Donnelly, Roth, and Baker, *Operation Just Cause*, p. 226.

children were held as "hostages" by the Americans in four houses along Freedom road during the firefight, which prevented Giumale from using his 60mm mortars to bombard and destroy the American position around the Super 6-1 site during the night.[59]

Noncombatants complicated urban operations planning. For example, in Operation Just Cause, both American and Panamanian noncombatants were present. The families of U.S. soldiers stationed in Panama—as well as tens of thousands of other U.S. citizens throughout Panama City—needed to be secured and evacuated. Early planning for this contingency was called *Klondike Key*, also called a "noncombatant evacuation operation (NEO)."[60]    Often noncombatants appeared unexpectedly during the operation, and extra resources had to be diverted from the primary mission to take care of them.  At the Torrijos International Airport in OJC, the unexpected presence of 376 civilian airline passengers complicated the Ranger mission, resulting in several hostage crises.  During the Fort Cimmarron assault, dozens of Americans at the Caesar Park Marriott hotel were held hostage temporarily by PDF gunmen in civilian clothes.[61]

## Rules of Engagement

ROE influenced how military force was applied, which in turn determined friendly and noncombatant casualties.  Constructing and managing flexible ROE so that they were neither too restrictive nor too permissive was critical for a successful political-military strategy that targeted the will of the enemy.  In recent urban operations, balancing ROE proved to be difficult, especially in the high-intensity case.  When improper ROE resulted in excessive civilian deaths and collateral damage, other MOUT elements such as the media and enemy IO gained useful ammunition for their respective interests.  ROE

---

[59]U.S. officers disputed the notion that Somali mortars would have wiped out Task Force Ranger because U.S. anti-mortar radar and Little Bird gunships loitering overhead would have destroyed any mortar crew after firing one or two rounds. See Atkinson, "Night of a Thousand Casualties," p. A11.

[60]See Donnelly, Roth, and Baker, *Operation Just Cause*, p. 24.

[61]In most cases, the hostages were eventually released unharmed.  However, one unfortunate American, Raymond Dragseth, was executed with a bullet to the back of his head. See Donnelly, Roth, and Baker, *Operation Just Cause*, p. 230.

also affected tactics and prevented the use of armor, artillery, and airpower on occasion.

ROE tightened the connection between politics and military tactics. Clausewitz's famous statement that "war is merely the continuation of policy by other means" has even more relevance for urban operations because of the heavier political pressure inherent in MOUT.[62] As a result, MOUT tactics, techniques, and procedures (TTPs) sometimes conformed to a political logic more than a military logic (at least before excessive casualties began to occur).

On at least one occasion heavy-handed political considerations created a military disaster.[63] The balance between restrictive ROE and permissive ROE needed to be tailored to reduce noncombatant casualties and general human suffering yet also avoid compromising the safety of friendly forces. For the MOUT commander, an ROE tradeoff always existed: either restrict the use of airpower, artillery, and armor and accept higher infantry casualties as a result, or allow heavier weapons to inflict collateral damage and noncombatant casualties.

The problem of how to balance ROE was not new. Historically speaking, conventional forces in the past often started with restrictive ROE that prevented the use of heavy firepower, but were forced to relax the ROE once unacceptable numbers of friendly casualties were taken.[64] Chechnya continued that trend. Before the December 1994 assault into Grozny, the Russian defense minister, Pavel Grachev, promised that no tanks or artillery would participate in the attack. President Yeltsin announced on Russian TV that

---

[62]Clausewitz's dictum that military force is a means toward a political end appears to remain true. However, some authors argue that future opponents of the United States may not fight for political ends but for moral, religious, or existential ends. See Chapter 5 of Martin Van Creveld, *The Transformation of War*, New York: Free Press, 1991.

[63]The political demands for a quick victory in the Chechen War was a major reason why the initial assaults on Grozny were such a disaster. The rushed job led to poor planning, a commitment of undermanned and unready troops, and a reckless mechanized drive straight into the center of an ambush.

[64]Captain Kevin W. Brown makes this point, using Manila (1945), Seoul (1950), Hue City (1968), Panama City (1989), and Somalia (1993) as historical examples. See Captain Kevin W. Brown, "Urban Warfare Dilemma—U.S. Casualties vs. Collateral Damage," *Marine Corps Gazette*, Vol. 81, No. 1, 1997, pp. 38–40.

> For the sake of saving people's lives I have given instructions that bombing strikes which could lead to fatalities among the civilian population of Grozny be ruled out.[65]

Grachev later stated that "local inhabitants, taking advantage of the fact that servicemen could not use violence against the peaceful population, have been dragging [Russian] troops out of their vehicles."[66]  Restraint on the use of force was abandoned after unsupported infantry began taking heavy losses.  As one Russian general put it,

> They want me to fight without artillery and aviation.  So as to be humane.  But I can't send soldiers into battle like that!  Without preparing the ground for them.[67]

The Russians relaxed their ROE, allowing artillery and airpower to damage nearly every heavy building in Grozny (with the exception of some suburbs).  Grachev used more tanks because "there was no other way."[68]

Restrictive ROE lowered combat effectiveness, put lives in danger, and fostered a sense of frustration and lower morale.  The need for political restraint on the use of violence was easy for a scholar of Clausewitz to understand but less appreciated among teenage soldiers who were putting their lives in jeopardy.

If ROE stripped away key equipment and firepower, soldiers were forced to fight with unfamiliar tactics.  Restrictive ROE that kept units from using combined arms assault groups most likely caused more casualties.  Urban warfighters were trained to work in combined arms teams, usually with tanks or infantry fighting vehicles attached to infantry units.  In Somalia, the 10th Mountain Division learned upon its deployment that it could not use its artillery.[69]  Artillery

---

[65]Stated on December 27, 1994.

[66]Quote from Thomas, *The Caucasus Conflict and Russian Security*, p. 34.

[67]See Serdyukov, "General Pulikovskiy:  Fed Up!"

[68]"Russian Military Assesses Errors of Chechnya Campaign," *International Defense Review*, Vol. 28, Issue 4, April 1, 1995.

[69]LTC T. R. Milton, Jr., "Urban Operations:  Future War," *Military Review*, February 1994.

pieces often blast entry and exit holes for infantry to use, which avoids the use of doors and windows that may be booby trapped or covered by fire. ROE can also strip away close air support, attack helicopters, and many other crew-served weapons. These heavier weapons are useful not only for suppression and destruction of enemy strongpoints, but also for urban maneuver. AC-130 gunships were not available for close air support on October 3rd because their previous employment had resulted in too much collateral damage. As General Colin Powell put it, "they wrecked a few buildings and it was not the greatest imagery on CNN."[70]

In OJC, restrictive ROE sought to minimize collateral damage and noncombatant casualties by restricting the use of artillery and air-power. Only a field-grade officer could authorize indirect fire from mortars or howitzers. When civilians were present, the use of artillery, mortars, armed helicopters, AC-130 tube- or rocket-launched weapons, or M551 main guns was prohibited without the permission of a ground commander with at least the rank of lieutenant colonel. Close air support, white phosphorus, and incendiary weapons were also prohibited in areas containing civilians without approval from at least division level.[71] General Stiner himself controlled air strikes from fixed-wing aircraft.

Restrictive ROE also forced a change in infantry TTPs in Panama.[72] For example, troops were not allowed to blindly clear rooms with a grenade. Strict ROE hampered small-unit tactics in numerous ways. The SEAL disaster at Paitilla Airport has been blamed on ROE that prevented SEAL snipers from shooting the PDF sentries before the SEALs began their main assault on the hangar.

Permissive ROE escalated tensions on the ground, caused higher noncombatant casualties, eroded the support of the population, and made it more difficult to gather HUMINT. Chechnya demonstrated

---

[70]See Travis M. Allen, *Protecting Our Own: Fire Support in Urban Limited Warfare,* Carlisle Barracks, PA: U.S. Army War College, 1999, p. 22.

[71]Taw, *Operation Just Cause,* p. 24.

[72]It should be noted that in Panama, ROE varied according to the objective. One company of the 3/75 Rangers attacked the Torrios terminal under a "weapons tight" mode (cannot fire until fired upon), but they were "weapons free" when they assaulted *La Comandancia.*

how indiscriminate attack by artillery or airpower can be counter-productive. One of the most important objectives in the urban guerrilla conflict was the will of the indigenous population. Since the guerrillas relied on the indigenous population for concealment, sup-plies, and intelligence, peacemaking forces could only succeed if that support was cut off. Indiscriminate destruction strengthened the support of the population for the enemy.

The nature of support and stability operations (SSOs) demanded more flexible ROE. SSOs often involved complicated political goals that were subject to change as the operational environment shifted rapidly. For example, the fluid conditions in Somalia required that soldiers be given some ROE latitude. Somalia was a peacemaking environment characterized by civil war, poverty, and unemployment with large numbers of armed Somali males running around in "technicals."[73] Deadly force could be used when soldiers were fired on or when the enemy had "hostile intent."[74] ROE that allowed a "graduated response" to threats, like the ROE in Somalia, offered one type of flexible response.[75]

## Presence of the Media

Media presence was more significant during the past decade for sev-eral reasons. Both the number of reporters and the portability of their information technology increased. It was easier for reporters to

---

[73]Technicals were pickup trucks loaded with gunmen and/or crew-served weapons.

[74]The ROE for both UNITAF and UNOSOM II were initially the same. USCENTCOM developed ROE based on its standard peacetime ROE. The commander of the Joint Task Force (CJTF) was then allowed make ROE more restrictive but not more permis-sive by issuing operating rules based on the ROE to his forces. These rules were copied onto 35,000 unclassified cards and handed out to U.S. soldiers. These ROE also formed the basis of ROE for UNOSOM II. Coalition forces were responsible for their own conduct. Because UN military forces were assigned their own sectors of respon-sibility, there were no conflicts that involved different sets of ROE. In fact, most other nations did not pay as much attention to ROE and in many cases used U.S. ROE. See Lorenz, "Law and Anarchy in Somalia," p. 29; and Jonathan T. Dworken, "Rules of Engagement: Lessons from Restore Hope, *Military Review*, September 1994, p. 28.

[75]The lack of nonlethal weapons limited soldiers' ability to use a graduated response to provocation. Yelling and throwing rocks back at their tormentors was ineffective. Pepper spray later proved more useful, and the Somalis were eventually conditioned to back off when soldiers simply waved aerosol shaving cream cans. Colonel F. M. Lorenz, "Law and Anarchy in Somalia," *Parameters*, Winter 1993–1994, p. 34.

gain access to the fighting in peace enforcement missions.[76] Most belligerents found the media a useful information tool for PSYOP, IO in general, civil affairs, and public affairs. Recent operations reinforce the notion that a successful political-military strategy must take account of the media's influence.

The Mogadishu and Budyonnovsk[77] examples, in particular, demonstrated that shocking images of combat can sway public opinion in an open democratic society and create intense political pressure to cease hostilities, especially if the conflict does not involve vital national interests. The Somalis influenced American public opinion by providing the media with graphic images of a mutilated American corpse being dragged through the streets of Mogadishu. The broadcast of this brutal image turned out to be a pivotal event. The Chechen raid on Budyonnovsk was also a pivotal media event. Television images of screaming women and children turned the Russian rescue attempts to free hostages at the hospital into a public relations disaster that was transmitted around the world. The resulting public outcry generated enough political pressure for Yeltsin to order negotiations with the Chechens. Budyonnovsk led to the first cease-fire, which gave the Chechens time to regroup after the successful Russian operations of the spring of 1995.

The media's influence on information operations depended, of course, on the extent of its access to the battlefield. Access depended on the remoteness of the region and the nature of the mission. Humanitarian operations generally meant more media presence because of the standing agreement that the press have unlimited access. If the area of operations was remote and the mission was not humanitarian, media access could be controlled through the use of the press pool (which was effective in Grenada, Panama, and Operation Desert Storm).

In Panama, the media was effectively controlled during the first few crucial days of combat (no shocking images of war were released). The short notice and brief duration of the main fighting were the primary reasons for this, but the use of the pool system kept

---

[76]The more remote a battlefield is (like the Iraqi desert), the easier it is for the military to restrict and control reporters.

[77]See the PSYOP section for a detailed description of the fight at Budyonnovsk.

reporters off the battlefield.  Both in CONUS (continental United States) and Panama, preparations for Operation Just Cause were concealed well enough to maintain operational surprise.[78]  The Pentagon took some time assembling a press pool for OJC, and even when sixteen reporters finally did arrive in Panama, they were kept waiting in a parking lot until half of them gave up and returned home.  The Pentagon press pool arrived at Howard Air Force Base at dawn on D-day, but they were subjected to a lecture on Panamanian history and flown to Fort Amador where they witnessed a firepower demonstration.  They were not given access to combat infantrymen or wounded.[79]  The media center was poorly equipped, so the pool had difficulty in filing timely news reports.[80]  Most reporters holed up at the Quarry Heights Officer's Club and tried to share information on the fighting.[81]  Although a few reporters did skirt DoD's restrictive press pool system and managed to roam the streets on their own, no reporters covered the most intense fighting in Panama City or Colon during the first two days.[82]

Media access in the Chechen War stands in sharp contrast to Panama.  In general, the Russian military appeared to lack a cohesive strategy for controlling the media.  As one Russian commentator put it, the Army had a "weak contact and interaction with the mass media."[83]  Access to the Chechen War was so porous that one journalist

---

[78]For example, Cable News Network (CNN) had learned from past operations to watch Pope Air Force Base next to Fort Bragg for increased activity, a tip that the 82nd Airborne Division was getting ready to act.

[79]See McConnell, *Just Cause:  The Real Story of America's High-Tech Invasion of Panama*, p. 197. According to the Joint History Office, reporter requests to visit the troops were turned down due to a shortage of available helicopter transport.  See also Ronald M. Cole, *Operation Just Cause:  The Planning and Execution of Joint Operations in Panama, February 1988–January 1990*, Washington, D.C.:  Office of the Chairman of the Joint Chiefs of Staff, 1995, p. 48.

[80]See Pascale Combelles-Siegel, *The Troubled Path to the Pentagon's Rules on Media Access to the Battlefield:  Grenada to Today*, Carlisle Barracks, PA:  U.S. Army War College, 1996, p. 10.

[81]Some reporters were already present before the fighting erupted, but they were not given timely access to the dozens of battles raging across Panama. Donnelly, Roth, and Baker, *Operation Just Cause*, p. 411.

[82]See McConnell, *Just Cause:  The Real Story of America's High-Tech Invasion of Panama*, p. 2.

[83]See Zakharchul, "View of a Problem."

called it the "great drive-in war."[84]  The author of *Chechnya: Calamity in the Caucasus* was able to drive directly from Moscow to Grozny and interview Dudayev twice.

The Chechen War was the first war in which the Russian and foreign press were allowed to witness Russian combat operations.  It was also the first "TV war" for the Russian public.  Hundreds of reporters arrived in Chechnya as the tanks rolled in.  Not only did the usual Western press agencies cover the fighting, but ITAR-TASS (the semi-official Russian news source), NTV (Russia's biggest independent television station), and a gaggle of other Russian media types were in Chechnya.  Russian independent television stations regularly ran critical and embarrassing coverage.[85]

The official military press had a hard time keeping up with the civilian press because it was used to having privileged access.  At times the Russian military did try to influence and control the media's message, but this effort was minimal.[86]  When official Russian reports were released to the public, they often contradicted what the civilian media was reporting on the scene.  Official attempts to cover up casualties and downplay the carnage of the war often backfired when the truth was made available by the media.  Because the Russian disinformation campaign failed to account for the civilian media, it damaged soldier morale.  For example, during the August 1996 battle for Grozny, recorded radio messages between Russian soldiers fighting for their lives were released by a Russian news program:

> You telephone Moscow.  They are saying on the television it is an insignificant conflict.  What that really means is that we are surrounded and our checkpoint is being destroyed.[87]

In contrast to the Russians, Chechen IO used the media.  Dudayev gave nightly interviews to Radio Liberty.  Dudayev's storyline made it

---

[84]See Lieven, *Chechnya: Tombstone of Russian Power*, p. 119.

[85]The media appeared to have a pro-Dudayev position on many of the news items during this period.

[86]One example was the instance of a Russian television crew filming ferocious-looking Russian commandos firing automatic weapons into the smoking ruins of the presidential palace after it was taken.

[87]See Gall and de Waal, *Chechnya: Calamity in the Caucasus*, p. 338.

into the media while the Russian military's side did not.[88] The Chechens used mobile TV stations to override Russian TV transmissions and deliver messages from President Dudayev directly to the people. The Internet was used to raise funds from abroad and mobilize Russian pubic opinion against the war.[89]

The overall impact of the media on the outcome of the Chechen War is difficult to assess. Media reports and images generated both international and domestic political pressure, but the latter was by far the more influential.[90] Media coverage waxed and waned during the course of the war.

## PSYOP and Civil Affairs

PSYOP are actions to convey selected information to foreign audiences to influence their emotions, motives, objective reasoning, and, ultimately, their behavior.[91] Civil affairs support PSYOP because they establish, maintain, and improve relations between the military force and the civil authority and general population.

In all three of the case studies, PSYOP and civil affairs operations proved indispensable in influencing the will of the civilian populations involved. In Chechnya, PSYOP were used to increase the number of noncombatants, and they were conducted by combining media exposure with daring military raids into Russian cities. In Chechnya and Panama, PSYOP also proved effective against military forces with low morale and cohesion, respectively the Russian army and the PDF.

---

[88]See Thomas, "The Caucasus Conflict and Russian Security: The Russian Armed Forces Confront Chechnya, III. The Battle for Grozny, 1–26 January 1995," p. 89.

[89]See Thomas, "Some Asymmetric Lessons of Urban Combat."

[90]The destruction in Grozny (especially by airpower) raised an international protest by members of the OSCE, the Council of Europe, and the EU, resulting in several demarches and sanctions. See Baev, "Russia's Airpower in the Chechen War," p. 6.

[91]Psychological operations can be waged at all levels of war. Strategic PSYOP aim to influence the will of the civilian populations involved in the conflict. Operational and tactical PSYOP aim to erode the fighting will of the enemy forces and to induce their surrender, desertion, and defection, to bolster friendly morale, and to win or coerce support from local populations. See Hosmer, "The Information Revolution and Psychological Effects," p. 218.

The circumstances, duration, and nature of the specific conflict partly determined the influence that civilians had upon combat operations (and therefore the importance of PSYOP and CA). In Panama, civilians were ambivalent about the fighting, and the basic civil affairs mission for Restore Hope was to minimize civilian interference with military operations. In the Somalia peace operation, the nongovernmental organizations (NGOs) were the civil affairs experts and few military specialists were used. The U.S. military also remained aloof and conducted minimal PSYOP.[92] In Chechnya, the insurgent nature of the conflict ensured that PSYSOPs were conducted extensively by both sides. The will of the Chechen people and the Russian people—as well as the public opinion of the world—was at stake for both sides.

Operation Just Cause demonstrated how effective PSYOP and CA units are when they are used against an army with weak morale and poor support from the indigenous population.[93] During the initial combat operations, PSYOP personnel deployed with the infantry, carrying bullhorns and going from building to building to ask or demand the surrender of PDF holdouts.[94] Usually the PDF soldiers did surrender; at other times they offered token resistance or simply ran away. A combination of ROE and PSYOP that used a graduated response usually proved sufficient. At Fort Amador, a demonstration of 105mm cannon, .50 caliber machine guns, and antitank and small arms fire combined with loudspeaker countdowns induced a stream of prisoners out the rear of the threatened buildings. The city of Colon could have been a nasty urban fight but most of the PDF surrendered or fled. In the town of Coco Solo, a demonstration by two

---

[92]UNITAF had made some earlier efforts at PSYOP, including leaflet drops, deploying loudspeaker teams, running a radio station, and even producing a daily newspaper that contained verses of poetry from the Koran. Based on comments by Ambassador Robert Oakley at the RAND/TRADOC/MCWL/OSD Urban Operations Conference, Santa Monica, California, March 23, 2000.

[93]In Panama, American PSYOP and civil affairs personnel maneuvered with combat troops throughout combat operations. They also did so during the later support and stability operations in Restore Hope. Members of the 96th Civil Affairs Battalion and the 4th PSYOP Group were among the first U.S. forces to parachute into Panama.

[94]Civil affairs units were aided by the fact that many combat infantrymen spoke Spanish.

20mm Vulcan Gatling guns mounted on HMMWVs convinced the 8th Naval Infantry company to surrender at its PDF barracks.[95]

Hundreds of civil affairs troops eventually deployed to Panama to execute Promote Liberty, a civil affairs operation designed to control the population and prevent looting.  CA units restored basic functions throughout Panama City, established a police force, supervised the distribution of food, and even developed a grassroots organization to sell the new government to the Panamanian people.[96]  PSYOP teams focused on communication themes designed to quell further resistance—for example, that U.S. forces had only deployed to protect the lives and property of U.S. citizens, or that U.S. differences were with Noriega alone and not with the Panamanian people.[97]

The Chechens used PSYOP to maintain political pressure on Yeltsin's government to stop the conflict.[98]  Chechen PSYOP were effective because most Russian soldiers and civilians did not feel that vital national interests were at stake.  The Chechens knew it would be very difficult to actually destroy Russian armed forces in battle; they sought to destroy their opponent's will to fight.  The lack of political conviction and leadership on the Russian side created a vulnerability for Chechen PSYOP.  Since political support for Yeltsin's decision to invade Chechnya was weak from the start, both the Chechens and the Yeltsin administration understood that the will of the Russian people was an important target.  Moscow sought to bolster domestic

---

[95]See McConnell, *Just Cause:  The Real Story of America's High-Tech Invasion of Panama*, p. 152.

[96]Cole, *Operation Just Cause:  The Planning and Execution of Joint Operations in Panama*, p. 53.

[97]Ibid., pp. 53–54.

[98]Thomas writes that four types of PSYOP operations were employed in Grozny: intimidation, provocation, deception, and persuasion.  PSYOP were employed to change attitudes through either fear or anger.  Acts of intimidation ranged from the Chechen practice of stringing up Russian prisoners outside the windows of the Council of Ministers building so they could fire from behind them to Dudayev's threats to blow up nuclear reactors.  An example of provocation was the Chechen practice of firing on Russian helicopters from village centers, in order to provoke return fire.  Chechen villages and homes were invariably destroyed, further alienating the public.  Chechens also used many deception techniques such as dressing up as Russian soldiers or Red Cross workers.  Persuasion techniques included using loudspeakers and leaflets to talk the Chechens into surrendering their weapons.

support for the war by the use of disinformation about the types of weapons used against targets in civilian areas, friendly casualties, and noncombatant deaths.

The most effective PSYOP tools for the Chechens turned out to be the media and the use of dramatic surgical MOUT strikes into Russia. Two highly publicized Chechen raids into the Russian urban areas of Budyonnovsk and Pervomaiskoye garnered intense publicity about the conflict among the Russian people and the rest of the world.[99]

In the first raid, Chechen leader Shamil Basayev raided the Russian city of Budyonnovsk in June 1995 with 148 fighters, capturing a city hospital and taking several hundred hostages. In the ensuing drama, Basayev obtained a press conference (after executing 12 hostages) and paraded Russian women and children captives in front of television cameras. A botched rescue attempt by the Russians led to further civilian and military casualties, which subsequently led to negotiations between Basayev and Russian Prime Minister Chernomyrdin. The meeting was televised, which implicitly granted the Chechens official respect and recognition.[100] The Budyonnovsk raid helped to swing Russian popular opinion against the war, temporarily forced a cease-fire, and led to a round of peace talk negotiations.

The second raid occurred on January 1996, when Salman Raduyev led 250 men into the Russian province of Dagestan, attacked the city of Kizlyar, and seized about 3,000 hostages. After cutting a deal, the Chechen guerrillas loaded up several buses with hostages, and the whole group headed home to Chechnya. The column was stopped outside the village of Pervomaiskoye near the border. The Chechens dismounted and entrenched in the village, and the Russians gathered reinforcements over the course of the next five days. Eventually Russian tanks, helicopters, and artillery pounded Pervomaiskoye as infantry and the elite Alpha commandos fought their way forward into the village building by building. Chechen machine guns and RPGs were instrumental in beating back the undermanned Russian attack. After eight days of seesaw battle, the Russians decided to withdraw their infantry and pulverize the entire village with standoff

---

[99]See "Russia's War in Chechnya: Urban Warfare Lessons Learned 1994–96," p. 4.

[100]See Major Raymond C. Finch III, "A Face of Future Battle: Chechen Fighter Shamil Basayev," *Military Review*, June–July 1997.

fire. When the Chechens heard of the impending barrage (by listening in on Russian communications), they decided it was prudent to leave. They dispersed in groups of fifty, exfiltrating through the Russian lines with their hostages in tow. Most managed to escape back across the Chechen border.

Throughout the Pervomaiskoye crisis, Russian authorities attempted to cover up the excessive civilian and military casualties, but their efforts backfired when the media covered the brutal assault on the village and reported the truth to the Russian public.[101]   Bloody Russian civilians gave interviews about the disregard for innocent bystanders and savage lack of ROE. Dozens of Russians of all political persuasions publicly condemned the Yeltsin government, raising political pressure to finish the costly war.

The Chechens also used PSYOP to encourage civilians to migrate to the fighting. Captured Russian soldiers were shown on Russian TV, prompting the mothers of some to travel to Chechnya on their own and negotiate for their sons' lives.[102]  As one Russian mother said,

> Russian mothers are screaming at Yeltsin, telling him to stop, stop the war. He just doesn't care. He just stares like a ram at a new gate.[103]

They encouraged hundreds of Russian mothers living in Russia to launch a grassroots campaign to stop the war and save their sons who were prisoners.[104]  When the Chechens were holed up in the presidential palace, they called the mother of one Russian captive, Krayeva, and told her that "your son is with us. He is alive, and everything will be fine but you must demand an end to the war." Mrs.

---

[101]According to Gall and de Waal, overall casualties were 96 Chechen fighters killed, 26 hostages killed, and about 200 Russian military killed and wounded. See Gall and de Waal, *Chechnya: Calamity in the Caucasus,* p. 303.

[102]Director Sergei Bodrov's movie *Prisoner of the Mountains* (1997) was about two Russian soldiers held captive by the Chechens. The mother of one of the soldiers travels to Chechnya to beg for her son's life.

[103]See Steve Erlanger, "A More Confident Russia Presses Hard on Rebels," *The New York Times,* January 15, 1995.

[104]For example, when 50-odd Russian paratroopers were captured near the village of Alkhazurovo, the Chechens telephoned their mothers to come pick them up.

Krayeva organized meetings, sent letters and telegrams to Yeltsin, and eventually went to Grozny to beg for her son's life in person.[105]

The standard PSYOP methods the Russians used to target the Chechen population proved to be ineffective. Leaflets were dropped from Russian aircraft and loudspeakers attempted to convince the Chechen people to not support the guerrillas or fight. Chechen radio was jammed. Local television stations were destroyed. Part of the problem was lack of civil affairs units in the Russian army.[106]

## Political-Military Strategy

Recent operations demonstrated that a political-military strategy is necessary to coordinate all efforts—especially IO—to subdue the enemy will and sustain your own people's will. It was important to have clear objectives before using military force, to make sure benefits justify costs, to avoid mission creep, and to have a clear exit strategy.[107]

The Somali experience demonstrated the folly of ignoring this wisdom. It was not just the media images of dead Americans that prompted an eventual U.S. withdrawal—it was the combination of the images and the absence of clear national interests that caused the public outcry.[108] Peace operations in Somalia took place in an environment riddled with poverty, ethnic-cultural hatred, and anarchy. The Somalis did not follow war conventions. Under these

---

[105]She ended up running around the battlefield while under fire, her son dragged behind her on a litter. One report also indicates that a large group of Chechen women once appeared outside the presidential palace to plead for everyone to stop the bloodshed. Gall and de Waal, *Chechnya: Calamity in the Caucasus*, p. 214.

[106]See Thomas, "The Battle for Grozny: Deadly Classroom for Urban Combat," p. 91.

[107]At least one author has argued that an exit strategy is not necessary if no Americans are being killed. See John Mueller, "Public Opinion and Foreign Policy: The People's Common Sense," in Eugene R. Wittkopf and James McCormick (eds.), *The Domestic Sources of American Foreign Policy: Insights and Evidence*, Lanham, MD: Rowman & Littlefield Publishers, Inc., 1999, p. 57.

[108]Warren Strobel notes that the so-called CNN effect—which he defines as a loss of policy control on the part of policymakers because of the power of the media—seems to have an impact primarily when policy is weakly held, is already in the process of being changed, or is lacking public support. See Warren Strobel, "The CNN Effect," *American Journalism Review*, May 1996.

conditions, it may have been impossible to meet U.S. political goals (including a limit on casualties) given the military means available.

The political-military strategy in Operation Just Cause was well executed. As the operation began, President Bush immediately gave a moving speech to the American people to justify the invasion and rally public support. General Powell's ready access to both the President, the Secretary of Defense, and the State Department allowed him to provide detailed political-military guidance to his operational commanders. There was a high level of coordination between the decisionmakers in the White House Situation Room and the military commanders in the National Military Command Center. A Crisis Action Team worked with the support of Defense Intelligence Agency personnel in the Crisis Management Room to respond to political-military issues as they arose. Military officers in the CJCS-J-3 Conference Room met daily with the National Security Council and the State Department.[109]

In the Chechen War it appears that no coherent political-military strategy was followed. Even for a stoic people like the Russians who historically have always accepted high casualties in war, the linkage between political and military goals must be clear if they live in an open society where information on the costs of war is available.

The original decisionmaking body was the Security Council of the Russian Federation, which subsequently put Grachev in charge. The Russian political objective was to unseat Dudayev and replace him with a figurehead more compliant with Russia's political leadership. Grachev took charge of the Chechen operation himself after firing the entire top leadership of the NCMD[110] who initially commanded the botched operation. Since the Security Council and the Ministry of Defense ran the operation at the highest levels, it is unclear whether the Russian General Staff was in the loop and who

---

[109]Cole, *Operation Just Cause: The Planning and Execution of Joint Operations in Panama*, pp. 45–46.

[110]The NCMD (North Caucasus Military District) is the military district responsible for Chechnya. It borders four independent states: Ukraine, Kazakhstan, Georgia, and Azerbaijan.

influenced Yeltsin to decide on an invasion in the first place.[111] The fact that terror bombing of Grozny continued for two days after Yeltsin ordered it halted appears to confirm that Moscow's control over field commanders was weak.[112]

The lack of political leadership had a corrosive effect on the morale of the Russian army. Many soldiers had no idea why they were fighting.[113] Russian soldiers were especially bitter with Yeltsin and Grachev. As one sergeant put it,

> We are here to show that the man who runs Russia has real power. The empire is dead and nobody can face it. So we are here to show that Russia is still a great power. But every day we are here we show the opposite. I have never been in another war, so I don't know what morale was. But other soldiers fought to save Russia. We fight to save Yeltsin.[114]

Russia's political leaders did a poor job of communicating to the Russian public the national interests at stake in Chechnya. Because the political goals of the war were never clearly articulated and justified, discontent grew at home. The political leadership failed to mobilize public opinion in favor of the invasion, did not identify what the desired end state was, and had no exit strategy. The lack of a political-military strategy contributed to the Russian weakness that Chechens sought to exploit—an unwillingness to accept the costs of prolonged guerrilla warfare.

---

[111]See Steve Erlanger, "The World:  Behind the Chechnya Disaster; Leading Russia into the Quagmire," *The New York Times,* January 8, 1995.

[112]See Stephen J. Blank and Earl H. Tilford, *Russia's Invasion of Chechnya:  A Preliminary Assessment,* Carlisle Barracks, PA:  Strategic Studies Institute, U.S. Army War College, 1994, p. 8.

[113]Many writers have noted the increasing convergence between military and civilian social values in modern society.  Professional armies are more integrated into civilian life, with less separation and a corresponding lack of elitism among military men and women.  Physical standards are dropping, more women are assuming roles on the battlefield, and it is more difficult to isolate soldiers from the influences of mainstream culture.  Under these shifting conditions, individual soldiers demand to know clearly why they must put their lives on the line.

[114]Sergeant Vladimir Kalunin, quoted in Michael Spector, "The World; Killed in Chechnya: An Army's Pride," *The New York Times,* May 21, 1995.

In contrast, the Chechen will to fight was based on historical and cultural factors more than political factors.[115] In fact, most Chechens were not supporting Dudayev at the start of the war. It was only after the Russians started bombing Chechen homes and killing civilians that the public rallied behind Dudayev.

## IMPORTANT FACTORS THAT REMAIN FUNDAMENTALLY UNCHANGED

Many of the remaining elements of MOUT identified in *Modern Experience in City Combat* remained fundamentally unchanged in the 1990s (see Table 5). Defending a city like Grozny was still much easier when the attacker could not isolate it. MOUT was still characterized by nonlinear combat between infantry squads and platoons. Combined arms teams were still essential and their employment did not change. The effects of surprise and technology on urban operations were no more important in the last ten years than they were during World War II. Communication in urban operations was still hampered enough that situational awareness remained elusive. Situational awareness was improved, but soldiers continued to communicate and fight the same basic way their fathers did at Hue.

Airpower evolved, but it is unclear whether the change was efficacious in terms of combat outcomes. For example, the usefulness of airpower varied according to circumstance—aircraft and rotary craft were less than ideal against an infantry force armed with SAMs or RPGs and dispersed among noncombatants, while airpower was effective against identifiable strongpoints during clear weather.

The remainder of this study provides an explanation for this lack of fundamental change for the following elements: situational awareness and intelligence, airpower, technology, surprise, combined arms, and joint operations.

---

[115]Chechens are a distinct ethnic group (close kin to the Igush) with an elaborate system of customs. Their society and loyalties are based on the clan and village. They have fought the Russians since the reign of Catherine the Great in the late 18th century.

## Situational Awareness and Intelligence

Recent urban operations demonstrated that complete situational awareness remained an elusive goal, just as it was in the past.[116] There were two reasons for this in our case studies: the unavailability of HUMINT and an inability to transmit sufficient information in the harsh electromagnetic conditions of the urban landscape.

HUMINT was more effective than SIGINT in urban terrain, especially when many noncombatants were present.[117] Somalia was a classic example of this type of HUMINT-intensive environment. The commander of Task Force Ranger, Major General William Garrison, believed that the key to catching Aideed was timely intelligence provided by HUMINT. HUMINT came from interpreters, humanitarian agencies, NGOs, civil affairs, infantry, military police, and special operations forces units, and about 20 Somali agents for the CIA based in Mogadishu.[118]

Despite a technological advantage in C4ISR, conventional armies oftentimes did not enjoy superior situational awareness over more primitive armies because HUMINT was usually the most effective type of intelligence in a city filled with noncombatants.

With the support of the population and the intimate knowledge that comes from fighting in their own back yard, one can argue that clan leaders in Somalia knew as much about what was going on as the Rangers taking cover in their HMMWVs. Somali gunmen knew where U.S. servicemen were because they had the support of the

---

[116]In fact, complete situational awareness may never be possible. War is inherently chaotic. Clausewitz tried to describe the complexity and uncertainty of war as "friction." Friction is used to represent all the unforeseen and uncontrollable factors of battle. In other words, friction more or less corresponds to the factors that distinguish real war from war on paper. It includes the role of chance and how it slows movement, or sows confusion among various echelons of command, or makes something go wrong when it has worked a hundred times before. See Carl von Clausewitz, *On War*, Michael Howard and Peter Paret (ed. and trans.), New York: Knopf, 1993.

[117]Low-intensity urban warfare places renewed emphasis on human intelligence. See Milton, "Urban Operations: Future War," p. 43.

[118]Information obtained by bribing was of questionable reliability. The main intelligence failure turned out to be an underestimation of Aideed's firepower, particularly regarding the stockpiles of hundreds of RPGs and the threat they posed to helicopters. See Everson, *Standing at the Gates of the City*, p. 36.

indigenous population. Somali women and children walked right up to American positions during the firefight, pointing them out for hidden gunmen. Gunmen also concealed their locations by hiding in crowds of noncombatants.

In contrast, U.S. situational awareness was relatively poor. Pockets of Rangers and Delta Force commandos holed up in adjacent buildings were often unaware that friendly units were close by. Officers circling above in command helicopters had access to real-time video of the firefight, but the video did not properly communicate the raw terror and desperation of the situation on the ground.[119]

Conventional armies also relied primarily on wireless communication for their C4ISR, which suffered severe degradation in the urban environment. Signals were blocked and degraded by channel obstructions and the interference of radio traffic. Radio signals were absorbed and reflected by buildings, materials, and other electromagnetic traffic.[120]

Since the urban operations relied on infantry, man-portable radios were essential. Unfortunately, man-portable radios had severe power limitations and were often unreliable. In Somalia, the man-portable PRC-77 radios (with secure devices attached) inside convoy vehicles were incapable of establishing a link, so that some vehicles became separated from their convoys during the firefight.[121] In Chechnya, a shortage of portable battery chargers hampered man-portable communications and forced the Russians to rely on radios in infantry fighting vehicles.[122]

---

[119]The commanders in the TOC could see what was happening from their real-time videos beaming down to them from the Navy Orion plane circling over Mogadishu. See Kent DeLong and Steven Tuckey, *Mogadishu! Heroism and Tragedy*, Westport, CT: Praeger, 1994, p. 95.

[120]Russian observers in Chechnya noted that wireless radios in the VHF/UHF range were best. Transmitters should ideally be placed in basements and antennas placed on the roof or in windows facing the receiver, connected using coaxial cable. Ground and airborne relays were also used.

[121]See Captain Mark A. B. Hollis, "Platoon Under Fire," *Infantry*, January–February 1998.

[122]See Thomas, "Some Asymmetric Lessons of Urban Combat."

Grozny's urban terrain kept the Russians from establishing continuous command and control. Clear lines of sight were difficult to maintain. The tactics of urban warfare—small infantry teams using raids and ambushes to advance and maneuver along separated axes—often resulted in the isolation of a "main body" and a nonlinear deployment of troops.[123] The complex nature of three-dimensional urban terrain meant that radio links could change at any time, both when the unit remained stationary and when it moved. Command and control positions had to be chosen with care with respect to these electromagnetic and tactical considerations; despite the best planning, an element of uncertainty always underlay communications in the city.

To enhance their communication links, the Russians learned to amplify their signals by locating transmitters and receivers along routes where radio waves could "excite" buildings or reflect off them.[124] Some structures actually increased the strength of wireless transmission by acting as reradiators.

Sometimes a minimal communications profile in the urban environment could, in fact, bestow advantages. It was difficult for Russian EW assets to cut off Chechen communications because of the loose and unstructured command and control system the Chechens used. As one Chechen put it, "When there is shooting we just find each other."[125] The Somalis also used a primitive but effective form of communication. The SNA communicated by using human runners, by beating on 55-gallon drums, and by flashing lights across the city (their Motorola radios were surely jammed by U.S. electronic warfare assets). For communications during the course of the October 3 firefight, the Somali leader in charge, Giumale, avoided using

---

[123]See Lt. General Miron Pavlishin, "Multifunctional Communication Systems," *Armeyskiy Sbornik*, translated in FBIS FTS19950502000749, May 1996.

[124]Evidently the Russians did this in Berlin and Koenigsberg during the Great Patriotic War. For example, a radio signal could be sent along a street to bounce off a stone building at an intersection in order to communicate with a receiver located on a perpendicular street. In this way, buildings acted as passive relays. See Colonel Vitaliy Kudashov and Major Yuriy Malashenko, "Communications in a City," *Armeyskiy Sbornik*, translated in FBIS FTS19970502000659, January 1, 1996.

[125]See Spector, "Commuting Warriors in Chechnya."

cell phones and instead used written messages and human couriers to issue his commands.

In Panama, the conventional force did enjoy excellent situational awareness and intelligence, but this was due to very unique circumstances. The U.S. troops already stationed in Panama trained beforehand on the very terrain they were to fight over.[126] Units reconnoitered the actual routes they were assigned for OJC. A couple of units ended up fighting where they used to play volleyball or golf. Familiar terrain eased the psychological stress of combat and reduced the uncertainty inherent in the planning of any military exercise. U.S. soldiers knew how long it took to fly a helicopter from one objective to another; they knew what the lighting was like around the neighborhoods they needed to secure.[127] They knew which PDF units were likely to remain loyal.[128]

In general, locating people in urban terrain was, and will probably remain, a difficult task. U.S. space and air assets such as unmanned aerial vehicles (UAVs), satellites, high-altitude aircraft, and battle-management aircraft like JSTARS are limited in their ability to detect dismounted forces in urban terrain because of the technological limitations of sensors, the presence of noncombatants, the nature of low-intensity warfare, and other uncontrollable factors such as inclement weather. Noriega's success in eluding U.S. attempts to capture him was embarrassing.[129] U.S. intelligence faced similar

---

[126]Extensive training and planning occurred over the long buildup of tensions. Various exercises and rehearsals were planned and carried out by the Joint Special Operations Task Force (JSOTF). JTF Panama ran a series of exercises throughout the summer and fall of 1989, known as PURPLE STORMs and SANDFLEAs.

[127]See Donnelly, Roth, and Baker, *Operation Just Cause*, p. 167.

[128]During the earlier October 1989 coup attempt, intelligence was gathered on which Panamanian units were most mobile and loyal to Noriega, including the PDF 4th Infantry Company and the Battalion 2000.

[129]Despite a round-the-clock "Noriega" watch by SOUTHCOM in the weeks prior to invasion, the human and signals intelligence assets devoted to fixing Noriega's position failed to keep up with the wily leader. Noriega moved every four hours, routinely split his convoys, and used other deception techniques to keep his whereabouts unknown. U.S. HUMINT was poor in Panama. Loyal sources had not been developed and databases of local individuals were not up to date. Taw, *Operation Just Cause*, p. 18.

problems hunting for Aideed in Mogadishu.[130]  On one raid, the Americans accidentally seized a key UN ally and members of the UN development program.

The Russians had an equally difficult time tracking dismounted infantry in the urban environment.[131]  Chechen infantry continued to elude Russian forces throughout the war.[132]  Every time a Russian task force of mechanized forces and paratroopers managed to encircle a Chechen village, most Chechens were able to exfiltrate through the surrounding Russian units.

Situational awareness was also made more difficult when both sides dressed alike or when noncombatants wore attire similar to that of soldiers.  In Chechnya, both sides wore civilian clothes or old Russian pattern camouflage and other items of military dress.  Russian units used nonstandard uniforms, especially elite outfits that affected a "Rambo" look.  Some Russians were forced to buy civilian clothes because of supply problems.  In Somalia, males over the age of twelve were armed.  It was hard to tell if a Somali was a bandit or a hired security guard for a humanitarian relief organization (HRO).[133]

---

[130]The capture of specific individuals was difficult because individual Somalis looked very similar to the untrained eye.

[131]The most dramatic exception was the assassination of President Dudayev by pinpoint missile attack in April 1995.  Supposedly the missile homed in on Dudayev's satellite telephone.  See Baev, "Russia's Airpower in the Chechen War," pp. 5, 13.

[132]In general, Russian SIGINT was poor in Chechnya.  Chechen situational awareness was enhanced because of the poor communications security practiced by the Russians.  The rebels basically listened to Russian communications that were transmitted in the clear.  The Chechens were even able to deceive Russian aircraft into attacking their own people on occasion.  No mockups of Grozny were completed. Reconnaissance was poor, and Chechen strongpoints were not uncovered prior to the assault.  Maps were obsolete.  Russian officers relied on 1:50,000 and 1:100,000 scale maps because they lacked more appropriate 1:25,000 or 1:12,500 scale maps.  See "Russia's War in Chechnya: Urban Warfare Lessons Learned 1994–96," p. 7.  Routes of advance were not properly reconnoitered.  Cuts in funding before the war meant that many satellites were turned off, and few aerial photography missions were conducted prior to the invasion.  Russian HUMINT was also poor.  Not a single Chechen fighter was captured prior to the assault in Grozny.

[133]See Lorenz, "Law and Anarchy in Somalia," p. 28.

## Airpower

Airpower proved to be a mixed blessing in recent urban operations because of the presence of noncombatants, ROE, and capable air defense threats. Urban terrain, poor weather, and an inability to precisely engage dispersed infantry with air-to-ground munitions also contributed to the mixed performance of airpower.

On the positive side, airpower was effective in joint operations around the perimeter of small villages and towns that could be isolated, against specific strongpoints that could be pinpointed, and in open areas in clear weather. Attack helicopters provided security and route reconnaissance, overwatch, and suppressive fire for ground forces. For example, in Mogadishu, close air support from AH-6 gunships, Cobras, and Blackhawks was very valuable.[134] Attack helicopters also had a positive psychological effect for friendly troops. The mere presence of helicopters served as a deterrent, causing crowds and vehicles to disperse. As one Ranger reported, "Those helicopters saved us. The brass casings came down around us like rain."[135]

On the negative side, in general, airpower was not discriminate enough in the presence of noncombatants. Indiscriminate killing of noncombatants had adverse consequences for PSYOP, CA, and PA. For example, on September 10, 1993, SNA militia intermingled freely with hundreds of other Somalis, including women and children, as they swarmed against some UN peacekeepers who were attempting to clear a roadblock. In the ensuing battle, Cobra gunships succeeded in dispersing the attackers but killed about 100 Somalis, including noncombatants.[136]

Helicopters also appeared to be vulnerable in MOUT environments where dismounted infantry carrying man-portable SAM weapons

---

[134]The AH-1's TOW missiles and AIM-1 laser-designated 20mm cannon (boresighted to the gun) reduced collateral damage enough that they were able to place fires within 50 meters of friendly forces. See Jones, *Attack Helicopter Operations in Urban Terrain*, p. 43, and U.S. Army Forces, Somalia, *10th Mountain Division After Action Report*, Executive Summary, p. 43.

[135]See DeLong and Tuckey, *Mogadishu! Heroism and Tragedy*, p. 95.

[136]See David, "The United States in Somalia: The Limits of Power," p. 9.

could conceal themselves within crowds of noncombatants.[137] During the Mogadishu firefight, two helicopters were shot down and three were damaged and forced to retire.[138] The vulnerability of helicopters to ground RPG fire complicated the mission when Task Force Ranger was ordered to try to locate, secure, and defend the two helicopter crash sites for 15 hours. Extending the duration of the surgical strike no doubt added to the high number of casualties. Helicopters were also vulnerable to snipers.

Airpower was effective in Operation Just Cause because the opponent failed to mount a credible air defense capability.[139] In fact, so many aircraft were used in OJC, air traffic control turned out to be a big challenge.[140] The fire support provided by Spectre AC-130 gunships and Apaches suppressed strongholds like *La Comandancia* so that infantry, light armor, and mechanized infantry units tightened a noose around the PDF.[141] AH-64A Apache helicopters armed with Hellfire missiles were also introduced for the first time in OJC.[142]

Airpower enabled ground troops to conduct rapid maneuver when the terrain was sufficiently open and no serious air defense threat

---

[137]The vulnerability of rotary-wing aircraft is growing because of the proliferation of MANPADS and millimeter-wave (MMW) tracking radar-guidance systems for short-range surface-to-air missiles (SAMs) and anti-aircraft artillery (AAA). Most currently deployed radar-warning receivers cannot detect MMW signals. According to the Army, MANPADS "are, and will continue to be, the most lethal threat" and are currently in the inventories of 115 armed forces, terrorists, and drug traffickers and are widely available on the international arms market. See Bryan Bender, "Threat to Helicopters Is Growing," *Jane's Defence Weekly*, February 10, 1999. Sometimes tactics can be adjusted to reduce helicopter vulnerability. One Russian technique was to use captured high-rise buildings as cover and "pop up" to engage targets such as snipers and other weapons located in upper-story floors. See Celestan, *Wounded Bear*.

[138]Overall, three U.S. Blackhawk helicopters were shot down by RPG fire in Somalia. A QRF Black Hawk was shot down on August 25, 1993.

[139]Of the special operations aircraft in Panama (including the MH-47D, AH-6/MH-6, and UH-60A helicopters), 30 percent were damaged and three were shot down, including the AH-6 carrying American civilian Kurt Muse. See Taw, *Operation Just Cause*, p. 21.

[140]With up to 250 helicopters and airplanes flying around at night under blackout conditions, the airspace above Panama City became packed and dangerous.

[141]The actual damage caused by the airpower was minimal, though. For example, the Rangers who finally seized *La Comandancia* reported that the bottom floor remained intact. See Donnelly, Roth, and Baker, *Operation Just Cause*, p. 159.

[142]The Apache night-fighting capability was particularly useful.

materialized. At the operational level in Panama, helicopters were indispensable in maneuvering troops between the multiple operational targets. Airmobile and airborne methods of insertion were the preferred means of deployment given the lack of a real SAM or counter-air threat.[143] However, the more urbanized terrain in Mogadishu was an example of where aircraft were useful for inserting ground troops but not for extracting them. Landing zones large enough for helicopters were rare, and ground convoys were necessary to extract troops.

Russian airpower filled many crucial roles in Chechnya but was not a decisive element.[144] Even though Russian airpower did succeed in establishing control of the air by striking three key Chechen airfields outside Grozny and effectively destroying the Chechen air force, it could not provide effective CAS to friendly troops. Poor weather, the presence of noncombatants and ROE, and an inability to engage dismounted troops in urban terrain limited the effectiveness of airpower. For example, during the initial assault on Grozny, poor weather severely limited the employment of precision weapons.[145] As Benjamin Lambeth put it,

> the weather took a turn for the worse, confronting VVS [Russian air force] aircrews with blowing snow, severe icing, and heavy cloud buildup with a low ceiling and tops above 15,000 feet. This made

---

[143]There were reports that the PDF possibly possessed SA-7 or SA-14 surface-to-air missiles (SAMs), but these weapons did not materialize on the battlefield. U.S. airpower basically operated with impunity. The only serious PDF air defense force to see battle was at Rio Hato, where three ZPU-4 antiaircraft guns and several VF-150 and V-300 armored cars guarded the airfield. The air defense weapons were potentially dangerous given their ability to fire from four 14.5mm barrels into the transport aircraft carrying the airborne assault force. Fortunately, some Apaches used their 30mm chain guns to take out two ZPUs and a Spectre used its 105mm howitzer to take out the third.

[144]Russian aircraft included the SU-27 fighter-bombers and SU-25 attack aircraft. Mi-8 and Mi-24 helicopters were also used offensively throughout the operation. Strategic bombers included the MiG-31, Su-27, Su-25, Su-17, and Su-24 short-range bombers to strike Chechen targets such as bridges, oil facilities, ammunition dumps, and C2 facilities. Tu-22M3 long-range bombers were also used. See "Russia's War in Chechnya: Urban Warfare Lessons Learned 1994–96," p. 4.

[145]Pilots sometimes used poor weather as an excuse. Many times, military pilots refused to fly into areas where Spetsnaz were fighting by claiming that the weather was too poor. See Oleg Blotskiy, "Chechnya: A War of Professionals," *Nezavisimaya Gazeta*, translated in FBIS FTS19960822000828, August 22, 1996.

both high- and low-angle manual bombing impossible and also precluded any resort to electro-optical or laser guided weapons. Instead, the VVS was forced to use Su-24 Fencers in day and night level bomb releases from medium altitude (15,000–20,000 feet) against radar offset points or in inertial bombing against geographic coordinates, through heavy cloud cover. The gross inaccuracy of these deliveries resulted in many Russian losses to friendly fire.[146]

Russian ROE, when they were in effect, limited the combat employment of air-to-ground munitions. Russian pilots were ordered to avoid the destruction of residences and the killing of civilians.[147] ROE were eventually violated because of the limited supply of precision-guided weapons, poor weather, and a lack of training. This resulted in heavy civilian casualties.[148]

A notable success for Russian airpower was the bombing of the most potent symbol of Chechen resistance, the presidential palace, during the first battle for Grozny. Six Su-25s dropped BetAB 3,000-pound concrete-piercing bombs on the palace on January 17, 1995.[149] Two of the bombs penetrated the structure from top to bottom, leaving most of the surviving Chechens in shock.[150] Eyewitness accounts imply that the Chechens decided to evacuate after the Russians demonstrated they could penetrate down to the basement with air-delivered weapons.[151] When a specific strongpoint with a concentrated mass of Chechens could be identified, airpower proved effective.

Russian airpower also enjoyed more success during March–April 1995 when the weather improved, more ground observers were em-

---

[146]See Benjamin S. Lambeth, *Russia's Air Power at the Crossroads*, Santa Monica, CA: RAND, MR-623-AF, 1996, p. 201.

[147]See Aleksandr Borisov, "Viewpoint: This Is Not Afghanistan, the Climate Here Is Different," *Armeyskiy Sbornik*, translated in FBIS FTS19970523001807, August 1, 1995.

[148]Artillery caused most of the damage in Grozny. One press account estimates that between 10,000 and 40,000 civilians were killed by August 1995. See "The Casualties of Chechnya," *The New York Times*, August 10, 1995.

[149]See Baev, "Russia's Airpower in the Chechen War."

[150]See Lambeth, *Russia's Air Power at the Crossroads*, p. 202.

[151]See Alessandra Stanley, "Chechen Palace, Symbol to Rebels, Falls to Russians," *The New York Times*, January 20, 1995.

ployed for intelligence gathering, and combat operations shifted from Grozny to more open areas.[152] Su-24s carrying laser-guided bombs like the KAB-1500 helped Russian ground troops to capture the Chechen strongholds of Argun, Gudermes, and Shali in March and April. Vedeno and Shatoi also fell to a combination of armor and airpower with very few Russian casualties. Since these smaller villages and towns could be encircled, defense proved impossible.

The Chechens fielded the most robust air defense threat in this study.[153] Russian helicopters were vulnerable to improvised Chechen tactical air defense weapons such as truck-mounted 23mm cannons and 12.7mm heavy machine guns (the Chechens put both machine guns and mortars on 4×4 civilian SUVs and trucks). At least one Russian helicopter was shot down by an RPG. Four helicopters were shot down from mid-December 1994 to the end of February 1995; by May 1996, a total of 14 were lost and 30 damaged. Several more were shot down later during the final battle for Grozny. As a result, the Russians used helicopters mainly for noncombat missions.[154] The official line from General Vitaliy Pavlov was that helicopters were not suited for urban combat.[155]

Airpower turned out to be a poor PSYOP weapon. The Russians used air strikes to pressure local populations to stop supporting Chechen guerrillas and to make separate truces with the Russian-installed client government in Grozny. As one source notes, "Bombardments were both indiscriminate and discriminate: indiscriminate in that they were intended to kill and terrorize the civilian population, but

---

[152]The Chechens showed their respect for Russian airpower by their aggressive attempts to hunt down Russian forward air controllers whenever they located them.

[153]Chechen air defenses included the Strela-10 (SA-13) SAM and the Igla-1 (SA-16), four mobile SU-23/4 radar and optically tracked antiaircraft guns, six ZU-23 and DShK optically sighted machine guns, and possibly some U.S.-made Stinger SAMs. See Lambeth, *Russia's Air Power at the Crossroads*, p. 196; also see "Russian Military Assesses Errors of Chechnya Campaign," *International Defense Review*, Vol. 28, Issue 4, April 1, 1995. Lieven believes the Stinger rumor is pure speculation. He also believes that most Russian helicopters were shot down with heavy machine guns, not SAMs. See Lieven, *Chechnya: Tombstone of Russian Power*, p. 278.

[154]According to the Commander of Russian Army Aviation, General Pavlov, normally 65 to 70 percent of helicopter resources are used for combat (assaults, convoy scout, CAS), but in Chechnya only 17 percent were used for combat missions.

[155]This was problematic because Russian doctrine called for a top-down approach to capturing buildings, which required troops to be airmobiled onto rooftops.

discriminate in that they were sporadic and limited."[156]  In some cases whole villages were deliberately destroyed to punish a local separatist.  Russian aircraft also intentionally made low supersonic passes over Grozny, laying down sonic booms to simulate bomb explosions and intimidate the Chechens.[157]  Indiscriminate bombing losses eroded the support of the indigenous population and domestic support back home.

Since weather had a significant influence on the application of air-power, its influence on urban operations should be noted here.  In the surgical and precision MOUT cases, weather was not a factor due to the short duration of the conflicts (though in Panama city, fog and darkness in the early morning of December 20th made it a little more difficult for air assault troops to reach their multiple objectives).  In the prolonged high-intensity case of Chechnya, however, bad weather severely limited air operations during the initial assault on Grozny.[158]  Because of the limited capabilities of the radar and night-vision equipment on Russian helicopters, 95 percent of the days in February 1995 were listed as "nonflying days."[159]  The frequent appearance of rain and fog over the battlefield limited the use of air-delivered munitions.

## Technology

Urban warfare technologies employed in the 1990s did not differ significantly from technologies available before 1982.  Weapons re-mained essentially the same, especially when ROE prohibited the stronger side from fielding advanced tanks and artillery. Commercial-off-the-shelf (COTS) equipment, nonlethal weapons, and PGMs were either not used, not considered, or were not deci-

---

[156]Quote from Anatol Lieven, "The World Turned Upside Down:  Military Lessons of the Chechen War," *Armed Forces International*, August 1998, p. 43.

[157]See Lambeth, *Russia's Air Power at the Crossroads*, p. 202.

[158]For a helpful explanation of why Russian leaders decided to initiate the invasion during the worst weather of the year, see Finch, *Why the Russian Military Failed in Chechnya.*

[159]Helicopters were used only when visibility was 1,500 meters.

sive.[160]  Small arms weapons continued to decide the course of MOUT for the most part—in fact, some of the most effective technologies continued to be the sniper, the flame-thrower, and the rocket-propelled grenade (RPG).

For example, well-concealed snipers could still pin down entire formations of soldiers because no effective anti-sniper weapon had appeared on the battlefield.  Snipers were used extensively by the Chechens, including 30 female snipers from the Baltic.[161]  They operated from rooftops and from deep within upper-floor apartments, making them difficult to spot.  Snipers created a disproportionate psychological stress among the enemy.  As one man put it:

> During the entire time I spent in central Grozny in January 1995, whenever I was in the open I imagined the sights of a sniper's rifle zeroing in on my head from some high building half a mile away.[162]

Superior technology was oftentimes negated by ingenious countermeasures.  For example, the SNA were barefoot yet managed to keep up with Americans in their HMMWVs and helicopters because of their use of swarm tactics and roadblocks.  The Somali gunmen were on foot but were able to keep up because U.S. convoys were forced to fight from ambush to ambush.[163]

Technological improvisation was often useful.  The Russians improvised their equipment according to the circumstances.  Fine wire mesh screens and cages—which stood out about 25 centimeters from hull armor—were added to vehicles to guard against Molotov cocktails and the shaped-charge jets of molten material from

---

[160]Precision munitions were generally not used.  Russian high-precision artillery such as the 1K113 Smelchak (fired from Tiulpan 240mm mortars) and Santimetr guided projectiles (fired from 152mm howitzers) with laser target-indication and range finding were never used during the campaign.  Some authors speculate that Russian commanders did not want to "waste" these expensive munitions on Chechnya.

[161]See Kostyuchenko, "Grozny's Lessons."

[162]See Lieven, "The World Turned Upside Down," p. 40.

[163]Gunmen ran along streets parallel to the convoy, keeping up because the two five-ton trucks and six HMMWVs were stopping and then darting across intersections one at a time.  This gave the gunmen time to get to the next street and set up to fire at each vehicle as it came through.

RPGs.[164] Steel plates were installed along the sides and above the roof of engine and transmission compartments. Infantry were protected by hanging vertical blinds of canvas or blankets to block sniper fire around certain areas.[165] The Chechens used tarpaulins to cover vehicle view ports when they attacked them. None of these technologies are new.

Commercial-off-the-shelf (COTS) technology has yet to make a significant impact in urban operations. Both the Chechens and the Somalis possibly used cellular phones, but they were easily jammed or tracked during the significant firefights.[166]

Nonlethal technologies would have been quite useful in all three cases but were generally not available. Americans used some pepper spray in Mogadishu and the Russians found tear gas and smoke (including formulations containing white phosphorus) to be useful.[167]

## Surprise

The advantage of surprise was critical to the outcome of all three case studies, but it was neither more nor less decisive than in the past. At both of the critical turning points of the Chechen War—the initial disaster in Grozny in December 1994 and the Chechen counterattack in Grozny in March 1996—Russian commanders and soldiers alike were shocked by the strength of the Chechen resistance.[168] The

---

[164]Also, reshetka armor screens—which resembled a set of venetian blinds fabricated out of steel bars—were added to trap incoming RPG rounds. See Sergey Leonenko, "Capturing a City," *Armeyskiy Sbornik*, No. 3, translated in FBIS, March 1995.

[165]See Oleg Namsarayev, "Sweeping Built-Up Areas," *Armeyskiy Sbornik*, translated in FBIS FTS19970423002215, May 4, 1995.

[166]Apparently Chechen bands may have also carried one hand-held Motorola radio per eight-man team. Comments by Arthur Speyer, RAND/TRADOC/MCWL/OSD Urban Operations Conference, Santa Monica, California, March 22, 2000.

[167]White phosphorus is not prohibited under international war conventions. The Russians discovered that a lengthy inhalation of WP (20–30 minutes) caused severe irritation of the mucous membranes of the eyes, pharynx, and larynx. Protective mask filters could not block WP. See Leonenko, "Capturing a City."

[168]In contrast, Russian ground forces did not attempt to achieve surprise. From the beginning, their strategy was to produce a show of force—a.k.a. 1968 Prague style—by

Americans achieved operational surprise in OJC, positioned as they already were in Panama.[169] Task Force Ranger lost the element of surprise in Mogadishu because the Somalis knew the basic pattern that U.S. forces followed from previous raids. As one of Aideed's lieutenants would say, "The Americans already had done basically the same thing six times."[170] Each time a raid was conducted, the Delta commandos flew in to seize a target building, the Rangers would ring the target for security and helicopters would loiter to provide fire as needed.[171]

## Combined Arms (Infantry with Armor and Artillery)

All three cases reinforce current doctrine that combined arms teams are essential if you need to minimize friendly casualties. Armor lacked infantry support in Grozny and infantry lacked armor support in Mogadishu. Neither force fared well. At the same time, the use of combined arms teams resulted in more collateral damage and non-combatant casualties. This is why ROE sometimes prohibited their use.

Clearly ROE that prohibited the use of combined arms teams increased the risk in urban combat. In the surgical and precision cases, combined arms teams were generally restricted by ROE. No artillery or U.S. tanks were involved in Mogadishu, while the heaviest weapon in Panama was the ground- or air-based 105mm howitzer. In the high-intensity case, Russian artillery provided most of the firepower that destroyed Grozny and completed the seizure of the city.[172] Once

---

rolling a seemingly invincible armored force straight into the heart of Grozny to intimidate the Chechens into surrendering.

[169]American forces achieved tactical surprise at the early objectives. Later assaults, such as Paitilla and Rio Hato, were obviously compromised by the violence of ongoing fighting.

[170]Quoted from Col. Ali Aden in Atkinson, "The Raid That Went Wrong."

[171]The Americans did try to vary their tactics somewhat. Sometimes they went in by helicopter; sometimes they went in by truck. Sometimes they came out on aircraft; sometimes they came out on trucks. The basic template was the same, however. See Atkinson, "The Raid That Went Wrong."

[172]As is typical in any war, Chechen artillery and mortars inflicted the greatest number of Russian casualties during the initial fight for Grozny. See Gregory J. Celestan, "Red Storm: The Russian Artillery in Chechnya," Field Artillery, January–February 1997.

ROE were relaxed and collateral damage was allowed, artillery was used to flatten any strongpoint that impeded progress. This politically damaging approach was actually Soviet standard practice in World War II.[173] Direct fire destroyed most of the Chechen strongpoints, typically from a range of 150–200 meters.[174]

## Joint Operations

Joint operations occurred in all three cases, usually involving air and ground forces. For the most part, joint operations did not make a significantly greater impact compared to urban operations before 1982.

Most Russian operations in Chechnya were joint in nature by default because units from the Ministry of Interior, the Ministry of Defense, and the Federal Counterintelligence Service fought side by side. During OJC planning, most of the U.S. armed services got to participate. The Navy was given an opportunity to use SEALs for missions other than covert reconnaissance, and the Marines were ordered to assault the PDF in the vicinity of Howard Air Force Base. In Somalia, Navy SEALs and C3I assets were under Army control.

Command, control, and communication problems continued to plague joint operations. Communication between air and ground forces was a problem in all three case studies. In Panama, apparently, there was a communications failure at Paitilla Airport—the SEALs were not able to call in fire support from the Spectre gunship circling above.[175] During the attack on *La Comandancia*, poor situational awareness and communication possibly caused a Spectre to fire on U.S. troops, wounding twenty-one men.[176] In Mogadishu, naval reconnaissance aircraft had no direct line of communication

---

[173]According to one source, Russian artillery was responsible for destroying 80–90 percent of enemy targets in the "tactical zone" in World War II. See Celestan, *Wounded Bear.*

[174]See Celestan, *Wounded Bear.*

[175]See McConnell, *Just Cause: The Real Story of America's High-Tech Invasion of Panama*, p. 66.

[176]Two Rangers were also killed by friendly fire at Rio Hato. See Donnelly, Roth, and Baker, *Operation Just Cause*, p. 153.

with the convoys on the ground.[177]  Army attempts to guide the wandering line of vehicles toward the helicopter crash sites failed because of the delay in relaying directions to the ground commander.

In Chechnya, coordination was weak between light ground forces and aviation units.  Russian command and control was never unified in Chechnya—no joint headquarters existed in Moscow where operations could be coordinated by one commander.  As a result, poor lines of communication between the various services caused many cases of fratricide.  At one point, a Ministry of the Interior regiment fought a six-hour battle with an army regiment.[178]  In addition, the troops of the Ministry of Internal Affairs (MVD) were not designed, equipped, or organized for large-scale combat operations.  They normally never trained with regular army troops and they possessed no organic armor or artillery.[179]  It was difficult to integrate these police units into joint operations with the army.[180]  Miscommunication between Russian ground units and CAS assets also caused many cases of fratricide.

---

[177]The Orion pilots were not allowed to communicate directly with the convoy.  Their orders were to relay all communications to the Joint Operations Center (JOC) back at the beach.  Also, no direct radio communications existed between the Delta Force ground commander and the Ranger ground commander.

[178]See Celestan, *Wounded Bear*, p. 10.

[179]See Celestan, *Wounded Bear*.

[180]See lesson 7, "Russia's War in Chechnya: Urban Warfare Lessons Learned 1994–96," p. 3.

# Chapter Four

# CONCLUSIONS

The manipulation of information is becoming more central to urban operations because of recent technological, political, and social developments. For example, the media is more capable of transmitting battlefield video footage to civilian populations. War is now waged on humanitarian grounds. The U.S. public expects new precision weapons to inflict fewer casualties on civilians. Because of developments such as these, the support of the civilian populations involved in the conflict is even more critical. Noncombatants can conceal the enemy, provide intelligence, and be killed in front of a camera; in effect, they seriously complicate both the tactical and strategic environments. Increasingly, the enemy's will to fight can be influenced by civil affairs, public affairs, PSYOP, management of the media, balanced ROE, and information operations in general.

Recent lessons from Panama, Somalia, and Chechnya provide a snapshot of how these information-related factors work:

- The presence of noncombatants significantly affected tactics, planning, ROE, and political-military strategy. Noncombatants were present in greater numbers, they played an active role in the fighting, they made ROE more restrictive, and they attracted the media.

- Balancing ROE proved to be difficult, especially in the high-intensity case. Constructing and managing flexible ROE so that they were neither restrictive nor permissive was critical. When improper ROE resulted in excessive civilian deaths and collateral damage, other MOUT elements such as the media and enemy IO could exploit the damage for their own interests. ROE also affected tactics and prevented the use of armor, artillery, and air-power on occasion. As a result, MOUT tactics, techniques, and procedures (TTPs) sometimes conformed more to a political logic than to a military logic (at least before excessive casualties begin to occur).

- All belligerents found the media a useful information tool for PSYOP, IO in general, civil affairs, and public affairs.

- PSYOP and civil affairs operations proved indispensable in influencing the will of the civilian populations involved. PSYOP were used to increase the number of noncombatants. PSYOP were conducted by combining daring military raids with media exposure.

- The failure of political leadership to communicate the national interests at stake in Somalia and Chechnya lowered the public's threshold for casualties. It was important to have clear objectives before using military force, to avoid mission creep, and have a clear exit strategy. The lack of political leadership also had a corrosive effect on morale in the Chechnya case.

At the same time, the more "traditional" elements of MOUT—airpower, combined arms, situational awareness, and technology—remain crucial to the outcome of urban battle. In most cases, defeating the will of the enemy is still best accomplished by killing the enemy. In the last decade, tanks, artillery, and infantry performed this basic role quite well (albeit under more restrictive political constraints), as they have done since World War II. Traditional factors did not, however, change in any fundamental way in the three urban operations looked at here.

Significant technological improvements in urban operations may be possible in the future. If improvements can be made in the areas of precision fire and C3I, then the use of military force in urban operations can evolve into a much more flexible option (even in the face of severe political constraints). For example, the discriminate application of force in urban operations could be improved with systems capable of selectively engaging individuals in a crowd. As some authors note, there are ways to make urban operations more "precise."[1] Nonlethal tactics and technology offer some promise for handling noncombatants.[2]  On the C3I side, a better man-portable wireless radio might enable dismounted infantry to establish situational awareness between, and possibly inside, buildings.[3]  Air-to-ground joint communications could be better integrated.

Yet new weapons, equipment, and tactical adjustments are only part of the solution.  What is needed, as this case analysis has hopefully shown, is a more comprehensive approach that recognizes the increasing significance of informational elements—the media, ROE, noncombatants, PSYOP, PA, and CA.  More important, linkages between these factors—the political-military campaign plan—must target the will of the people.  In this age of restricted warfare, the effort to subdue the will of the enemy requires a systems approach that combines information-related activities with the application of military force.

For example, an aggressive information campaign by the White House could help shape public (and congressional) opinion about what constitutes a vital interest and what does not.  Resources could be directed toward perception management as well as precision weapons. Aggressive intelligence efforts to dig out proof of criminal or hostile actions by the enemy could help demonize them in the eye of the public.  Public affairs activities could help raise the ceiling on how many casualties a public is willing to tolerate.  Information operations targeted at the indigenous population in the theater can

---

[1]See Major Charles Preysler, *Going Downtown: The Need for Precision MOUT*, Fort Leavenworth, KS:  School of Advanced Military Studies, U.S. Army Command and General Staff College, 1994, p. 38.

[2]For example, Rangers used harmless flash-bang grenades to disperse noncombatants from combatants so they could avoid killing unarmed people.

[3]One goal of DARPA's Small Unit Operations program is to develop a mobile wireless communication system for widely dispersed tactical units.  This equipment will be capable of supporting a tactical internet based on dismounted soldier and mounted vehicle nodes without having to rely on a fixed ground infrastructure, essentially a "communication on the move" capability.  The most promising type of system would be a mobile mesh network of communication nodes that are able to buffer, store, and route packets of information.  The main component would be the software radio—a packet switching, non-line-of-sight radio that uses software applications to perform some of the major communications functions that analog components do in current radios.

help win support and improve HUMINT. Nonlethal weapons, appropriate ROE, and PSYOP could help control noncombatants at the tactical level. All of these actions would complement the use of American military force and influence an enemy's will to fight.

Detailed recommendations for IO cells operating on future battlefields are outside the scope of this monograph. The intention here is to simply highlight the changing relevance of information-related activities in urban operations.

In future conflicts, it should be anticipated that some U.S. adversaries will recognize the growing importance of these information elements and leverage them as part of an asymmetric response to American firepower. War has always been waged in both the physical and the informational realm, but the political, technological, and social changes under way today make it imperative that we pay more attention to the latter.

## Books, Articles, Manuscripts, RAND Reports

Allard, Kenneth, *Somalia Operations: Lessons Learned,* Washington, D.C.: Fort McNair, National Defense University Press, 1995.

Allen, Travis M., *Protecting Our Own: Fire Support in Urban Limited Warfare,* Carlisle Barracks, PA: U.S. Army War College, 1999.

Arquilla, John, and Theodore Karasik, "Chechnya: A Glimpse of Future Conflict?" *Studies in Conflict and Terrorism,* Vol. 22, No. 3, July–September 1999.

———, and David Ronfeldt (eds.), *In Athena's Camp: Preparing for Conflict in the Information Age,* Santa Monica, CA: RAND, MR-880-OSD/RC, 1997.

Atkinson, Rick, "The Raid That Went Wrong; How an Elite U.S. Force Failed in Somalia," *Washington Post,* January 30, 1994.

———, "Night of a Thousand Casualties; Battle Triggered U.S. Decision to Withdraw from Somalia," *Washington Post,* January 31, 1994.

Baev, Pavel K., "Russia's Airpower in the Chechen War: Denial, Punishment and Defeat," *Journal of Slavic Military Studies,* Vol. 10, No. 2, London: Frank Cass, June 1997.

Bender, Bryan, "Threat to Helicopters Is Growing," *Jane's Defence Weekly,* February 10, 1999.

Bennett, Scott D., and Allan C. Stam III, "The Declining Advantages of Democracy: A Combined Model of War Outcomes and Duration," *Journal of Conflict Resolution*, Vol. 42, No. 3, June 1998.

Blank, Stephen J., and Earl H. Tilford, *Russia's Invasion of Chechnya: A Preliminary Assessment*, Carlisle Barracks, PA: Strategic Studies Institute, U.S. Army War College, 1994.

Blotskiy, Oleg, "Chechnya: A War of Professionals," *Nezavisimaya Gazeta*, translated in FBIS FTS19960822000828, August 22, 1996.

Borisov, Aleksandr, "Viewpoint: This Is Not Afghanistan, the Climate Here Is Different," *Armeyskiy Sbornik*, translated in FBIS FTS19970523001807, August 1, 1995.

Bowden, Mark, "Blackhawk Down: An American War Story," *The Philadelphia Inquirer*, November 16–December 16, 1997.

——, *Blackhawk Down: A Story of Modern War*, New York: Atlantic Monthly Press, 1999.

Braestrup, Peter, *Big Story: How the American Press and Television Reported and Interpreted the Crisis of Tet 1968 in Vietnam and Washington*, Boulder, CO: Westview Press, 1977.

Brown, Captain Kevin W., "Urban Warfare Dilemma—U.S. Casualties vs. Collateral Damage," *Marine Corps Gazette*, Vol. 81, No. 1, January 1997, pp. 38–40.

Byman, Daniel L., Matthew C. Waxman, and Eric Larson, *Air Power as a Coercive Instrument*, Santa Monica, CA: RAND, MR-1061-AF, 1999.

Celestan, Major Gregory J., "Red Storm: The Russian Artillery in Chechnya," *Field Artillery*, January–February 1997, pp. 42–45.

——, *Wounded Bear: The Ongoing Russian Military Operation in Chechnya*, Fort Leavenworth, KS: U.S. Army, Foreign Military Studies Office, August 1996.

Clausewitz, Carl von, *On War*, Michael Howard and Peter Paret (ed. and trans.), New York: Knopf, 1993.

Cole, Ronald M., *Operation Just Cause: The Planning and Execution of Joint Operations in Panama, February 1988–January 1990,* Washington, D.C.: Office of the Chairman of the Joint Chiefs of Staff, 1995.

Combelles-Siegel, Pascale, *The Troubled Path to the Pentagon's Rules on Media Access to the Battlefield: Grenada to Today,* Carlisle Barracks, PA: U.S. Army War College, 1996.

Cushman, John H., "Death Toll About 300 in October 3 U.S.-Somali Battle," *The New York Times,* October 16, 1993.

David, Colonel William C., "The United States in Somalia: The Limits of Power," *Viewpoints,* 95-6, June 1995, located at http://www.pitt.edu.

DeLong, Kent, and Steven Tuckey, *Mogadishu! Heroism and Tragedy,* Westport, CT: Praeger, 1994.

Dick, Charles J., "A Bear Without Claws: The Russian Army in the 1990s," *Journal of Slavic Military Studies,* Vol. 10, No. 1, London: Frank Cass, March 1997.

Donnelly, Thomas, Margaret Roth, and Caleb Baker, *Operation Just Cause: The Storming of Panama,* New York: Lexington Books, 1991.

Dworken, Jonathan T., "Rules of Engagement: Lessons from Restore Hope," *Military Review,* September 1994.

Erlanger, Steve, "A More Confident Russia Presses Hard on Rebels," *The New York Times,* January 15, 1995.

———, "Russian Troops Take Last Chechen Cities," *The New York Times,* April 1, 1995.

———, "The World: Behind the Chechnya Disaster; Leading Russia into the Quagmire," *The New York Times,* January 8, 1995.

Everson, Major Robert E., *Standing at the Gates of the City: Operational Level Actions and Urban Warfare,* Fort Leavenworth, KS: School of Advanced Military Studies, U.S. Army Command and General Staff College, 1995.

Finch, Major Raymond C., III, "A Face of Future Battle: Chechen Fighter Shamil Basayev," *Military Review*, June–July 1997.

————, *Why the Russian Military Failed in Chechnya*, Fort Leavenworth, KS: U.S. Army, Foreign Military Studies Office, 1998.

Gall, Carlotta, and Thomas de Waal, *Chechnya: Calamity in the Caucasus*, London and New York: New York University Press, 1998.

Gartner, Scott Sigmund, and Gary M. Segura, "War, Casualties, and Public Opinion," *Journal of Conflict Resolution*, Vol. 42, No. 3, June 1998, pp. 278–300.

Geibel, Adam, "Lessons in Urban Combat: Grozny, New Year's Eve, 1994," *Infantry*, Vol. 85, No. 6, November–December 1995, pp. 21–25.

Gerwehr, Scott, and Russell Glenn, *The Art of Darkness: Deception and Urban Operations*, Santa Monica, CA: RAND, MR-1132-A, 1999.

Glenn, Russell, *Combat in Hell*, Santa Monica, CA: RAND, MR-780-A/DARPA, 1996.

————, *Marching Under Darkening Skies: The American Military and the Impending Urban Operations Threat*, Santa Monica, CA: RAND, MR-1007-A, 1998.

————, *". . . we band of brothers": The Call for Joint Urban Operations Doctrine*, Santa Monica, CA: RAND, DB-270-JS/A, 1999.

Goligowski, Major Steven P., *Future Combat in Urban Terrain: Is FM 90-10 Still Relevant?* Fort Leavenworth, KS: School of Advanced Military Studies, U.S. Army Command and General Staff College, 1995.

————, *Operational Art and Military Operations on Urbanized Terrain*, Fort Leavenworth, KS: School of Advanced Military Studies, U.S. Army Command and General Staff College, 1995.

Graham, Bradley, "War Without 'Sacrifice' Worries Warriors," *The Washington Post*, June 29, 1999.

Grishin, Anatoli, "Accounting for the Chechen War," *Itogi*, September 24, 1996 (translated by Olya Oliker).

Groves, Brigadier General John R., "Operations in Urban Environments," *Military Review*, July–August 1998.

Holley, David, "Serbs Rally Around Their Leader," *Los Angeles Times*, March 26, 1999.

Hollis, Captain Mark A. B., "Platoon Under Fire," *Infantry*, January–February 1998.

Holsti, Ole R., *Public Opinion and American Foreign Policy*, Ann Arbor, MI: University of Michigan Press, 1996.

Hosmer, Steve T., *Constraints on U.S. Strategy in Third World Conflict*, Santa Monica, CA: RAND, R-3208-AF, 1985.

———, "The Information Revolution and Psychological Effects," in Zalmay M. Khalilzad and John P. White (eds.), *Strategic Appraisal: The Changing Role of Information in Warfare*, Santa Monica, CA: RAND, MR-1016-AF, 1999.

Jones, Major Timothy, *Attack Helicopter Operations in Urban Terrain*, Fort Leavenworth, KS: School of Advanced Military Studies, U.S. Army Command and General Staff College, December 1996.

Kempster, Norman, "Leaders and Scholars Clash Over Legality," *Los Angeles Times*, March 26, 1999.

Korbut, Andrei, "Learning by Battle," *Nezavisimoye Voyennoye Obozreniye*, December 24, 1999 (translated by Olya Oliker).

Kostyuchenko, Aleksandr, "Grozny's Lessons," *Armeyskiy Sbornik*, translated in FBIS FTS1995110100633, November 1, 1995.

Kudashov, Colonel Vitaliy, and Major Yuriy Malashenko, "Communications in a City," *Armeyskiy Sbornik*, translated in FBIS FTS19970502000659, January 1, 1996.

Lambeth, Benjamin S., *Russia's Air Power at the Crossroads*, Santa Monica, CA: RAND, MR-623-AF, 1996.

Larson, Eric, *Casualties and Consensus: The Historical Role of Casualties in Domestic Support for U.S. Military Operations*, Santa Monica, CA: RAND, MR-726-RC, 1996.

Leonenko, Sergey, "Capturing a City," *Armeyskiy Sbornik*, No. 3, translated in FBIS, March 1995.

Lieven, Anatol, *Chechnya: Tombstone of Russian Power*, New Haven and London: Yale University Press, 1998.

————, "The World Turned Upside Down: Military Lessons of the Chechen War," *Armed Forces International*, August 1998.

Lorell, Mark, and Charles Kelley, *Casualties, Public Opinion, and Presidential Policy During the Vietnam War*, Santa Monica, CA: RAND, R-3060-AF, 1985.

Lorenz, Colonel F. M., "Law and Anarchy in Somalia," *Parameters*, Winter 1993–1994.

————, "Standing Rules of Engagement: Rules to Live By," *Marine Corps Gazette*, February 1996.

McConnell, Malcolm, *Just Cause: The Real Story of America's High-Tech Invasion of Panama*, New York: St. Martin's Press, 1991.

McLaurin, R. D., Paul A. Jureidini, David S. McDonald, and Kurt J. Sellers, *Modern Experience in City Combat*, Aberdeen Proving Ground, MD: U.S. Army Human Engineering Laboratory, March 1987. Distributed by Defense Technical Information Center, Cameron Station, Alexandria, VA.

Miller, David, "Big City Blues," *International Defense Review*, Vol. 28, Issue 3, March 1, 1995.

Milton, LTC. T. R., Jr., "Urban Operations: Future War," *Military Review*, Vol. 74, No. 2, February 1994, pp. 37–46.

Mueller, John E., *Policy and Opinion in the Gulf War*, Chicago: University of Chicago Press, 1994.

————, *War, Presidents and Public Opinion*, New York: John Wiley, 1973.

Mukhin, Vladimir, and Aleksandr Yavorskiy, "War Was Lost Not by the Army, but by Politicians," *Osobaya papka OF nezavisimaya gazeta*, Internet edition, No 37 (2099), February 29, 2000.

Namsarayev, Oleg, "Sweeping Built-Up Areas," *Armeyskiy Sbornik*, translated in FBIS FTS19970423002215, May 4, 1995.

Naylor, Sean D., "A Lack of City Smarts? War Game Shows Future Army Unprepared for Urban Fighting," *Army Times*, May 11, 1998.

Netherly, Major Phillip T., *Current MOUT Doctrine and Its Adequacy for Today's Army*, Fort Leavenworth, KS: School of Advanced Military Studies, U.S. Army Command and General Staff College, 1997.

Newell, Major Mark R., "Tactical-Level Public Affairs and Information Operations," *Military Review*, December 1998–February 1999.

Novichkov, N. N., et al., *Rossiiskie Vooruzhennye Sily v Chechenskom Konflikte: Analiz, Itogi, Vyvody*, Paris, Moscow: Kholveg-Infoglob, Trivola, 1995.

Nye, Joseph S., Jr., and William A. Owens, "America's Information Edge," *Foreign Affairs*, Vol. 75, No. 2, March/April 1996.

Pavlishin, Lt. General Miron, "Multifunctional Communication Systems," *Armeyskiy Sbornik*, translated in FBIS FTS19950502000749, May 1996.

Preysler, Major Charles, *Going Downtown: The Need for Precision MOUT*, Fort Leavenworth, KS: School of Advanced Military Studies, U.S. Army Command and General Staff College, 1994.

———, *MOUT Art: Operational Planning Considerations for MOUT*, Fort Leavenworth, KS: School of Advanced Military Studies, U.S. Army Command and General Staff College, 1995.

Raevsky, Andrei, "Russian Military Performance in Chechnya: An Initial Evaluation," *Journal of Slavic Military Studies*, Vol. 8, No. 4, London: Frank Cass, December 1995.

Rick, Charles, *The Military–News Media Relationship: Thinking Forward*, Carlisle Barracks, PA: U.S. Army War College, 1993.

Serdyukov, Mikhail, "General Pulikovskiy: Fed Up!" *Sobesednik,* translated in FBIS, September 1996.

Shapiro, Jeremy, "Information and War: Is It a Revolution?" in Zalmay M. Khalilzad and John P. White (eds.), *Strategic Appraisal: The Changing Role of Information in Warfare,* Santa Monica, CA: RAND, MR-1016-AF, 1999.

Spector, Michael, "Commuting Warriors in Chechnya," *The New York Times,* February 1, 1995.

———, "Jan. 15–21: In Grozny's Rubble; Seizing a Chechen Symbol, Russians Claim Victory, but the War Goes On," *The New York Times,* January 22, 1995.

———, "The World; Killed in Chechnya: An Army's Pride," *The New York Times,* May 21, 1995.

Stanley, Alessandra, "Chechen Palace, Symbol to Rebels, Falls to Russians," *The New York Times,* January 20, 1995.

Stearns, Captain Scott C., "Unit-Level Public Affairs Planning," *Military Review,* December 1998–February 1999.

Strobel, Warren, "The CNN Effect," *American Journalism Review,* May 1996.

Surozhtsev, "Legendary Army in Grozny," *Novoye Vremya,* No. 2–3, January 1995, pp. 14–15.

Taw, Jennifer, *Operation Just Cause: Lessons for Operations Other Than War,* Santa Monica, CA: RAND, MR-569-A, 1996.

———, and Bruce Hoffman, *The Urbanization of Insurgency: The Potential Challenge to U.S. Army Operations,* Santa Monica, CA: RAND, MR-398-A, September 1994.

———, and John E. Peters, *Operations Other Than War: Implications for the U.S. Army,* Santa Monica, CA: RAND, MR-566-A, 1995.

———, David Persselin, and Maren Leed, *Meeting Peace Operations' Requirements While Maintaining MTW Readiness,* Santa Monica, CA: RAND, MR-921-A, 1998.

Thomas, Timothy, *The Caucasus Conflict and Russian Security: The Russian Armed Forces Confront Chechnya, Part I and II,* Fort Leavenworth, KS: U.S. Army, Foreign Military Studies Office, 1995.

———, "The Caucasus Conflict and Russian Security: The Russian Armed Forces Confront Chechnya, III. The Battle for Grozny, 1–26 January 1995," *Journal of Slavic Military Studies,* London: Frank Cass, Vol. 10, No. 1, March 1997.

———, "Some Asymmetric Lessons of Urban Combat: The Battle of Grozny (1–20 January 1995)," unpublished draft manuscript, Fort Leavenworth, KS: U.S. Army, Foreign Military Studies Office, January 8, 1999.

———, "The Battle for Grozny: Deadly Classroom for Urban Combat," *Parameters,* Summer 1999.

Van Creveld, Martin, *The Transformation of War,* New York: Free Press, 1991.

Van Dyke, Carl, "Kabul to Grozny: A Critique of Soviet Counter-Insurgency Doctrine," *Journal of Slavic Military Studies,* Vol. 9, No. 4, London: Frank Cass, December 1996.

Waxman, Matthew C., "Siegecraft and Surrender: The Law and Strategy of Cities as Targets," *Virginia Journal of International Law,* Virginia Journal of International Law Association, Vol. 39, No. 2, Winter 1999.

Willis, G. E., "Remembering Mogadishu: Five Years After the Firefight in Somalia, Some Say U.S. Forces Abroad Still Are Reeling from It," *Army Times,* October 1998.

Wittkopf, Eugene R., and James McCormick (eds.), *The Domestic Sources of American Foreign Policy: Insights and Evidence,* Lanham, MD: Rowman & Littlefield Publishers, Inc., 1999,

Yin, Robert K., *Case Study Research,* Newbury Park, CA: Sage Publications, 1988.

Zakharchul, Colonel Mikhail, "View of a Problem," *Armyskiy Sbornik,* translated by FBIS, FTS19970423002216, March 28, 1995.

"The Casualties of Chechnya," *The New York Times,* August 10, 1995.

"The Chechen Conflict: No End of a Lesson?" *Jane's Intelligence Review—Special Report,* September 1, 1996.

"Russia Pounds Rebel Positions Outside Capital of Chechnya," *The New York Times,* May 21, 1995.

"Russian Military Assesses Errors of Chechnya Campaign," *International Defense Review,* Vol. 28, Issue 4, April 1, 1995.

"Russia's War in Chechnya: Urban Warfare Lessons Learned 1994–96," *Marine Corps Intelligence Activity Note,* CBRS Support Directorate (MCIA-1575-xxx-99), November 1998.

## Field Manuals, Official Government Documents

Chairman of the Joint Chiefs of Staff, *Joint Doctrine for Command and Control Warfare (C2W),* Joint Pub 3-13.1, 7 February 1996.

Chairman of the Joint Chiefs of Staff, *Joint Doctrine for Information Operations,* Joint Pub 3-13, 9 October 1998.

Chairman of the Joint Chiefs of Staff, *Doctrine for Joint Psychological Operations,* Joint Pub 3-53, 10 July 1996.

Department of the Army, *An Infantryman's Guide to Combat in Built-up Areas,* Field Manual (FM) 90-10-1, Washington, D.C.: U.S. Government Printing Office, 1993.

Department of the Army, *Military Operations on Urbanized Terrain,* Field Manual (FM) 90-10, Washington, D.C.: U.S. Government Printing Office, 1979.

Department of the Army, *Public Affairs Operations,* Field Manual (FM) 46-1, 30 May 1997.

Department of the Navy, *Military Operations on Urbanized Terrain (MOUT),* Marine Corps Warfighting Publication 3-35.3, Washington, D.C.: U.S. Government Printing Office, April 1998.

The White House, *A National Security Strategy for a New Century,* December 1999.